Elementary Reaction Kinetics

Third Edition

J.L. LATHAM
Ph.D., C.Chem., F.R.I.C.
University of Glasgow

and

A.E. BURGESS
Ph.D., C.Chem., M.R.I.C.
Glasgow College of Technology

BUTTERWORTHS
LONDON - BOSTON
Sydney - Wellington - Durban - Toronto

The Butterworth Group

United Kingdom London	Butterworth & Co (Publishers) Ltd 88 Kingsway, WC2B 6AB
Australia Sydney	Butterworths Pty Ltd 586 Pacific Highway, Chatswood, NSW 2067 Also at Melbourne, Brisbane, Adelaide and Perth
South Africa Durban	Butterworth & Co (South Africa) (Pty) Ltd 152–154 Gale Street
New Zealand Wellington	Butterworths of New Zealand Ltd T & W Young Building, 77–85 Customhouse Quay, 1, CPO Box 472
Canada Toronto	Butterworth & Co (Canada) Ltd 2265 Midland Avenue Scarborough, Ontario, M1P 4S1
USA Boston	Butterworth (Publishers) Inc 10 Tower Office Park, Woburn, Mass. 01801

First published 1962
 Reprinted 1964
 Reprinted 1968
Second edition 1969
 Reprinted 1976
Third edition 1977
 Reprinted 1980

ISBN 0 408 46102 0

© Butterworth & Co (Publishers) Ltd 1977

LIBRARY OF CONGRESS CATALOGING IN PUBLICATION DATA
Latham, Joseph Lionel.
  Elementary reaction kinetics.
  Bibliography: p.
  Includes index.
  1. Chemical reaction, Rate of. I. Burgess,
A.E., joint author. II. Title.
QD502.L38 1977   541'.39   77–902
ISBN 0-408-46102-0

Printed in England by Billing & Sons Limited,
Guildford, London and Worcester

Preface

The object of this book is to provide an introduction to the main ideas of chemical reaction kinetics. It is intended for students reading the subject for the first time, and should be particularly useful to the undergraduate in providing a less rigorous introduction to the subject than is found in specialist texts. The book, however, does deal with a wide range of topics with the intention of giving the reader a broadly based understanding of this rapidly developing subject area.

Thus the book should prove to be suitable not only for the student of chemistry, but also for students following courses in physical or life science who require an understanding of elementary reaction kinetics.

The first three chapters cover the concepts and terminology of chemical kinetics, the basic rate laws and an introduction to the methods of measuring reaction rates. Chapter 4 deals with the effect of temperature on reaction rate and the concept of activation energy. This leads into chapters on the basic theories of reaction rate, the special problem of unimolecular reactions and the kinetic investigation of reaction mechanisms.

The importance of catalysis is emphasized by separate chapters on heterogeneous, homogeneous and enzyme catalysis. The subject of chain reactions and free radical processes is treated in Chapter 11, and then the reader is introduced to the techniques used for studying fast reactions. Examples of the analysis of experimental results are given in Chapter 13. The concluding chapter has sections giving a summary of useful mathematics, the least-squares method of line fitting and an outline of the thermodynamic concept of chemical equilibrium which is fundamental to the theory of reaction rates.

To deal with the fact that reaction rates vary continuously with time, it is necessary to have some knowledge of the mathematical technique specially devized to deal with continuously varying quantities – namely infinitesimal calculus. To help the reader who cannot readily bring to mind the mathematical results required in this text, a compact summary of such results is given in Chapter 14. When quoted in the text, these results are referred to by the letter M (for Mathematical Result).

The most important equations developed in the text are placed in a rectangular box on the first occasion on which they occur. This has been done to help the reader to pick out the important points that can usefully be committed to memory.

The authors have felt it justifiable in certain cases to simplify (some might say over-simplify) complex topics. One of the failings of students of reaction kinetics is that they sometimes cannot see the wood for the trees! Many excellent specialist texts exist which deal with kinetic topics with thoroughness and mathematical rigour. These, however, tend to be somewhat overpowering for a beginner. It is hoped that this book will enable the reader to realize the scope of reaction kinetics, and will be a useful introduction to the specialist texts dealing with those topics for which greater sophistication is needed. Suggestions for additional reading are made at the end of the book.

The IUPAC recommendations for the units of physical quantities have been used throughout. In naming compounds classical or recommended names are used, depending on which seemed the more appropriate.

CONTENTS

LIST OF PRINCIPAL SYMBOLS

(i) Roman Letters

A	reactant species
A	pre-exponential factor of Arrhenius equation or Debye–Hückel constant
a	concentration of A
a_A	activity of A
B	reactant species
b	concentration of B
$C^{\ddagger}$	transition state species C
c	molecular velocity
$c^{\ddagger}$	concentration of transition state species C
d	differential operator
E	enzyme
E	energy or electromotive force
$E^{\ddagger}$	energy of activation
exp	exponential function, i.e. $\exp(x) = e^x$
F	mathematical function (general form)
G	Gibbs free energy or conductance
H	enthalpy
h	Planck's constant
I	integration constant
J	joule
K	kelvin
K	equilibrium constant
k	rate constant (forward reaction) or force constant
k_{-1}	rate constant (back reaction)
kJ	kilojoule
l	litre
log	logarithm (to the base 10)
ln	logarithm (to the base e)
M	molar mass
m	metre
m	molecular mass or proportionality constant
mol	mole
N	newton
N	Avogadro constant

n	molecular concentration
Pa	pascal (= 1 N m^{-2})
p	pressure
P	probability factor (in collision theory)
Q	ratio of actual to most probable velocity
R	gas constant
S	substrate or siemens
S	entropy
s	second
s	number of vibrational degrees of freedom
T	temperature
t	time
$t_{1/2}$	half-life
V	molar volume
X	experimental value
x	change in concentration due to reaction
y_A	activity coefficient of substance A, concentration basis
Z	collision constant
z	charge on an ion

(ii) Greek Letters

Δ	operator meaning 'final value of . . . minus initial value of . . .'
η	viscosity
θ	fraction of surface covered
μ	ionic strength, chemical potential
ν	frequency
π	circumference divided by diameter of circle
ρ	density
Σ	operator meaning in 'the sum of . . .'
σ	molecular collision diameter
τ	relaxation time

(iii) Miscellaneous

$\propto$	proportionality sign
∞	infinity sign
$\int$	integral sign
∂	partial differential operator
subscript o	refers to initial state
subscript -1	refers to back reaction
superscript $\ddagger$	refers to transition state
!	factorial sign
$[\cdots]$	molar concentration of . . .

INTRODUCTION

The subject of reaction kinetics is concerned with the detailed study of the rates of chemical reactions. The experimental part of the subject deals with ways of measuring precisely the variation of the concentrations of reacting substances with time. These measurements are carried out in such a way that the effects of temperature, pressure, catalysts, isotopes, radiation, etc. on the rate of reaction can be assessed.

Interpretation of the experimental results leads to a better understanding of the mechanism of reactions. The combination of the results of a large number of experiments gives rise to general theories of chemical reactivity.

From a fundamental point of view there are two important aspects to any process of change. The first is the extent of the change; the second is the time taken to accomplish the change. It is no consolation to the master of a ship which has struck an iceberg, to be told that if only he had waited a few years for equilibrium to be reached, the iceberg would have melted and the ship would have been safe!

In chemical reactions, the problem of predicting the products when equilibrium is eventually reached is dealt with in chemical thermodynamics. The equilibrium constant can in principle be predicted from free energies obtained from measurements made in calorimeters. When equilibrium is achieved rapidly, it is possible to predict quite accurately the relative proportions of the reactants and products in simple cases. Thermodynamics has, however, nothing to say about the *rate* at which a chemical reaction will occur. An analogy may be drawn with the case of a stone falling under gravity. Knowledge that its ultimate position will be on the ground (a 'thermodynamic' result) enables no deductions to be made about the rate of fall (a 'kinetic' result).

The limitation of thermodynamics in explaining chemical processes is seen by considering the reaction of hydrogen and oxygen at one atmosphere pressure and room temperature. The reaction appears to occur instantaneously when the mixture is sparked. On the other hand, in the absence of a spark or a catalyst, no reaction is detectable after several years. Thermodynamic calculations show, however, that the reaction is accompanied by a large decrease in free energy, and so should be capable of occurring spontaneously with some vigour. (The kinetic explanation of why one does not look for a gas leak with a lighted candle is discussed in Chapter 11.)

There are many examples of chemical reactions that occur at a measurable rate. It is worth noting that the process of life itself depends on the combined effect of many thousands of chemical reactions, each proceeding at a steady rate at body temperature. The dramatic effect of temperature on rate of reaction (see Chapter 4) is illustrated by the fact that a 10 °C rise in temperature of the human body invariably leads to death.

It is fairly easy to show experimentally that the rates of chemical reactions vary with time, but careful experiments are needed to show that this variation is regular and can be described by a mathematical equation. The first accurate kinetic study of a chemical reaction was carried out in 1850 by Wilhelmy, who measured the rate of conversion of an acidic solution of sucrose into glucose and fructose. This reaction was especially suitable for kinetic study as the amount of reaction could be found at any time by measuring the optical rotation of the solution in a polarimeter. Wilhelmy found that, at a given concentration of acid, the rate of reaction at any instant was proportional to the amount of sucrose remaining in solution.

In 1862, Berthelot and St. Gilles made a careful study of the equilibrium between acetic acid, ethyl alcohol, ethyl acetate and water:

$$CH_3CO_2H + C_2H_5OH \rightleftharpoons CH_3CO_2C_2H_5 + H_2O$$

They were able to show that in this reversible reaction the rate of the forward reaction was proportional to the concentration of ethyl alcohol multiplied by the concentration of acetic acid.

The idea that the rate of a chemical reaction at a given temperature depends on concentration was generalized by Guldberg and Waage who, in 1863, stated the *Law of Mass Action.* In modern terms this states that *the rate of a chemical reaction is proportional to the concentration of each reactant.* This law provides a quantitative basis for kinetic investigations.

One point that may cause confusion is that the Law of Mass Action is defined in some textbooks in terms of 'active mass'. This is for historical reasons as Guldberg and Waage used the term in their original publication. 'Active mass' has, however, no connection with the thermodynamic concept of 'activity'. In fact the rates of chemical reactions are proportional to the concentrations of the reagents rather than to their thermodynamic activities. To avoid confusion it is best to state the Law of Mass Action in terms of concentration or for gas reactions in terms of partial pressure*. Guldberg and Waage further showed that the position of chemical equilibrium can be explained quantitatively for a reversible reaction by assuming a *dynamic* rather than a *static* equilibrium, in which the rate of the forward reaction is equal to the rate of the reverse reaction.

The Law of Mass Action may be given a molecular interpretation when applied to a reaction of the type

$$A + B \rightarrow \text{products}$$

in which one molecule of A reacts with one molecule of B. Before the two molecules can react with one another to form a compound they must first meet. The number of collisions between the molecules of A and of B on simple probability theory is proportional to the number of molecules of A multiplied by the number of molecules of B, which in turn is proportional to the concentration of A multiplied by the concentration of B. The Law of Mass Action thus follows from the reasonable assumption that the number of molecules of A and B that react is proportional to the number of collisions between them.

In *one-stage reactions* the law of mass action may be applied directly to the concentrations of the reactants. If, however, a reaction occurs in a series of consecutive stages, the law must be applied successively to each individual stage of the reaction.

These simple ideas do not apply to all reactions, but they lead to the principle which is basic to the quantitative study of chemical reaction kinetics. This is that, at a given temperature and pressure, the reaction rate is a function of the concentrations of the reactants. It is this theme that is pursued in the following chapters of this book.

*It may be recalled that for a gas obeying the ideal gas law, the partial pressure is proportional to the molar concentration.

1
CONCEPTS AND TERMINOLOGY OF CHEMICAL KINETICS

The purpose of this chapter is to introduce the main concepts in reaction kinetics in such a way that they can be easily referred to later. These ideas are used and explained in more detail in the subsequent chapters.

1.1 Kinetics

The word *kinetic** was originally used to mean 'pertaining to motion'. For example, the kinetic theory of gases deals with properties of gases which are dependent on molecular motion. In chemical reactions there is no apparent motion, but there are changes in concentration. The phrase 'chemical kinetics', which is often abbreviated to 'kinetics', is used to describe the quantitative study of change in concentration or pressure with time brought about by chemical reaction.

1.2 Categories of Reactions

Chemical reactions can be divided into two broad categories: *homogeneous* and *heterogeneous*. In the former only one phase is present and the system is uniform in composition throughout. Reactions in a single solvent in which no solid catalyst is used are homogeneous. So, too, are gas reactions other than those occurring at a catalytic surface. Some examples of homogeneous reactions are:

$$H_2 + I_2 \rightarrow 2HI \text{ (in the gas phase)}$$

$$CH_3COCl + CH_3OH \rightarrow CH_3COOCH_3 + HCl \text{ (in the liquid phase)}$$

In heterogeneous reactions the mixture is not uniform throughout, and reaction occurs at phase boundaries. This is typically the case when solid catalysts are used, as in the following examples:

*This word is not restricted to the scientific field, e.g. kinetic art.

$$C_2H_5OH \xrightarrow{\text{alumina}} C_2H_4 + H_2O$$

$$2NH_3 \xrightarrow{\text{tungsten}} N_2 + 3H_2$$

Enzyme reactions have some of the features of both homogeneous and heterogeneous reactions.

1.3 Rate of Reaction

The precise meaning of the term 'rate of reaction' is not self-evident. A simple definition might be the mass of product formed in a given time under stated conditions. There are, however, two objections to this interpretation:

(a) the concentration of reactants changes as the reaction proceeds, and hence constant conditions cannot be maintained;

(b) the amount of product formed depends on the initial quantity of reactants as well as on their concentrations and chemical reactivities.

These difficulties can be overcome by defining the *rate of reaction* as the *decrease in concentration per unit time of one of the reactants.* Concentrations are usually expressed in moles per litre (written as mol 1^{-1}); the time in seconds.

This definition may be expressed more concisely using the notation of the differential calculus. If [A] represents the concentration of the reactant A, measured at time t, then the rate is defined as

$$\text{rate} = -d[A]/dt \qquad (1.1)$$

The minus sign occurs because the concentration of the reactant *decreases* with increasing time.

An alternative definition of the rate of reaction is in terms of the *product.* If a represents the initial concentration of A and x represents the concentration of product at time t then

$$\text{rate} = +dx/dt \qquad (1.2)$$

or

$$\text{rate} = \frac{-d(a - x)}{dt} \qquad (1.3)$$

In equation 1.2 the sign is positive since the concentration of products *increases* with time (see M18).

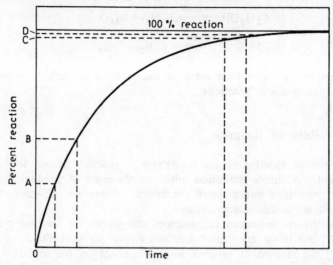

Figure 1.1   Variation of reaction rate with time

A typical curve of percentage reaction plotted against time is shown in *Figure 1.1*. It illustrates two general features of chemical kinetics. The first is that the amount of reaction in a given time interval decreases as the reaction proceeds. The amounts of reaction for two equal time intervals are shown on the graph. AB, which corresponds to the earlier time, is much greater than CD. The rate of reaction, using equation 1.2, is given by the slope of the curve, which clearly decreases with increasing time.

The second point is that there is no definite instant of time at which the reaction is completed, as the curve approaches 100% reaction asymptotically.

1.4   Rate Constant

The rate constant is a measure of the rate of a given chemical reaction under specified conditions. It may be defined in words as *the rate of change in concentration of reactant or product with time for a reaction in which all the reactants are at unit concentration.* The definition is helpful in that it gives some physical meaning to the rate constant.

As an example, in the decomposition of benzenediazonium chloride in water:

$$C_6H_5N_2Cl + H_2O \rightarrow C_6H_5OH + N_2 + HCl$$

the rate constant is 4.24×10^{-4} s^{-1} at $40.0\,^{\circ}C$. From the definition, in one litre of a one molar solution, benzenediazonium chloride will start to decompose at the rate of 4.24×10^{-4} mole per second at this temperature. It follows also that if the unit of time used in measuring the rate constant is changed, say, from seconds to minutes, the value of the rate constant will increase by a factor of sixty.

The above definition of rate constant cannot always be used quantitatively because:

(a) chemical reactions are not, in general, carried out with all the reactants at one mole per litre; indeed many reactants are not soluble to this extent;

(b) even if the system were initially at unit concentration, as soon as reaction occurred the concentration would alter and the reaction rate would change;

(c) rate constants are affected slightly by pressure, ionic strength, etc. (Chapter 5) which are here assumed to be constant.

In order to obtain a precise definition, applicable to all cases, the calculus notation must be used.

At constant temperature the rate of reaction depends upon the concentrations of the reactants, although it is not always directly proportional to them as would be implied by the law of mass action. In more formal language:

rate of reaction = a constant multiplied by a function of the concentrations of reactants

$$dx/dt = k \times F(a,b,c,\ldots) \tag{1.4}$$

where k is the rate constant and $a,b,c,\ldots$ represent the concentrations of the reactants A, B, C, ... at time t. The function $F(a,b,c,\ldots)$ represents some mathematical expression that is a characteristic of the reaction. This function only involves the products of the concentrations, each raised to a certain power, and so it becomes unity when all the reactants are at unit concentration since $1^n = 1$. Under these conditions, therefore, $dx/dt = k$, which is the rate constant in agreement with the earlier definition.

For any particular reaction, the value of k is *constant at a given temperature and pressure*, and is a convenient quantitative measure of chemical reactivity. It must be stressed, however, that

k increases rapidly with temperature, and so equations like 1.4 are only valid when the temperature is kept constant.

1.5 Rate Law

An equation like 1.4 which relates the rate of reaction to the concentrations of reactants is called the *rate law* or the *rate equation*. It is determined by experiment (see Chapter 3). In simple reactions the rate law takes one of the forms shown in

Table 1.1 Typical rate laws

Rate law	Order	
$\dfrac{dx}{dt} = k(a - x)^0 = k$	0	(1.5)
$\dfrac{dx}{dt} = k(a - x)$	1	(1.6)
$\dfrac{dx}{dt} = k(a - x)^2$	2	(1.7)
$\dfrac{dx}{dt} = k(a - x)(b - x)$	2	(1.8)
$\dfrac{dx}{dt} = k(a - x)(b - x)^2$	3	(1.9)

Table 1.1. In complex reactions the rate law often takes a more elaborate form and fractional powers may occur.

The rate laws 1.4–1.9 are differential equations and are referred to as the *differential form of the rate law*. They may be integrated (see Chapter 2) to give the rate law in a form into which experimental results may be substituted directly. The result of the integration is referred to as the *integrated form of the rate law*.

1.6 Order of Reaction

The word 'order' is used in pure mathematics in the classification of differential equations. The rate laws shown in *Table 1.1* are all differential equations. In chemical kinetics, these equations are classified according to the *order of reaction*.

The order is usually a small whole number, but in special cases it may have a fractional value or be zero. It is formally defined as *the sum of the powers of the concentration terms that occur in the differential form of the rate law.* Thus a reaction with rate law 1.6 is first order since the concentration term is raised to the power of unity. Rate laws 1.7 and 1.8 are second order and 1.9 is third order. Since any number to the power of zero equals one, rate law 1.5 may be written $dx/dt = k(a - x)^0$ and is, therefore, of zero order (see M6). Rate laws and corresponding orders are shown in *Table 1.1.*

It cannot be too strongly emphasized that the *order* of reaction is entirely an experimental quantity which can be measured without knowing the mechanism of the reaction. The order of the reaction is completely independent of the reaction stoichiometry. This follows from the fact that the order is determined solely by finding out which rate law best fits the experimental data. The order *cannot* be found merely by looking at the chemical equation for the reaction. Finally, it must not be confused with *molecularity*, which is defined below.

1.7 Molecularity of Reaction

The strength of a chain cannot exceed the strength of the weakest link. There is a corresponding principle in kinetics, namely that the overall rate of a process cannot exceed the rate of the slowest stage. This 'bottleneck principle' is frequently observed in everyday life. For instance, if a large crowd is leaving a building which has only a few small exits (such as a theatre or cinema), the time taken to empty the building is determined by the number of people who can squeeze through the doors per second. Whether people run or walk to the doors makes no difference to the rate of leaving; it merely affects the size of the queue inside.

If a chemical reaction proceeds in a series of sequential stages, then the rate of reaction is limited by the slowest stage. This stage is referred to as the *rate determining* or *rate controlling stage.* The molecularity of a reaction is defined as the number of molecules or ions that participate in the *rate determining stage.*

This definition is not quite precise, as in some cases it is not easy to explain the exact meaning of 'participate'. It is shown in Chapter 5 that all reactions pass through a transition state. If this is accepted, then the definition of molecularity is *the number of molecules or ions from which the transition state is formed.* The molecularity is a theoretical quantity in that to evaluate it the mechanism of reaction must be known or assumed.

In contrast to the order, the molecularity is necessarily a small whole number and cannot be zero or fractional. The terms 'unimolecular' and 'bimolecular' are used to describe reactions which have a molecularity of one and two, respectively.

In many cases the order and molecularity are equal, e.g. the thermal decomposition of t-butyl alcohol

$$t\text{-}C_4H_9OH \rightarrow C_4H_8 + H_2O$$

is a first-order unimolecular reaction. On the other hand, the reaction of iodine with acidified aqueous acetone

$$CH_3COCH_3 + I_2 \rightarrow CH_3COCH_2I + HI$$

is bimolecular and of zero order (see p. 76). The oxidation of nitric oxide

$$2NO + O_2 \rightarrow 2NO_2$$

is bimolecular and of third order (see p. 82).

It should be noted that some well established textbooks on kinetics do not distinguish clearly between order and molecularity and refer indiscriminately to second-order reactions as 'bimolecular'.

1.8 Half-life

In some cases it is convenient to define the rate of a chemical reaction by stating *the time taken for 50% reaction to occur.* This time is called the *half-life* (*symbol* $t_{1/2}$). It is widely used in describing the rates of radioactive decay in which the half-life is independent of the amount of material present (see equation 2.11). In other orders of reaction the half-life depends upon the initial concentrations as well as upon the rate constant.

If there is more than one reactant, then a precise value cannot be given to the half-life unless a particular reactant is stated. For example, in a bimolecular second-order reaction between A and B, if the initial concentrations are in the ratio of 3 to 2 then:

at the half-life of A, 75% of B has reacted;
at the half-life of B, 33.3% of A has reacted.

1.9 Infinite Time

Chemical reactions approach completion gradually (see *Figure 1.1*) and so there is no instant of time at which the reaction finishes. However, it is often necessary to know what the final concentrations of the reactants will be. By *infinite time* is meant *the time at which the reaction is complete for all practical purposes* (say more than 99.9% reaction). Thus one can tell experimentally when 'infinite time' has been reached by the constancy of the composition of the reaction mixture.

In first-order reactions, 99.9% reaction occurs in ten times the half-life, which can be taken as 'infinite time'. In second-order reactions a considerably longer period is needed. In many cases the time taken to reach 99.9% reaction can be conveniently reduced by warming a sample of the reaction mixture. This technique is often used when it is necessary to know the concentrations of the reactants at infinite time in order to calculate the rate constant.

It should be noted that the word 'infinite' is used in this context in a relative sense, and that infinite time can vary from a fraction of a second to many years, depending on the reaction.

1.10 Fast Reactions

The word *fast* is used in this context to describe chemical reactions having a half-life between a few microseconds and a few seconds. The rates of such reactions cannot be measured by sampling methods. Special techniques, such as those described in Chapter 12, are necessary.

2

REACTION RATE LAWS

2.1 The Need for Integration

The main problem dealt with in this chapter is that of converting the rate laws into a form from which the rate constant and order of reaction can be calculated. As mentioned in the discussion of equations 1.5–1.9, the rate laws for the different orders of reaction are differential equations since they all involve the term dx/dt for the rate of reaction (where x represents the concentration of product formed or reactant consumed at time t).

In practice, if sampling techniques are used dx/dt cannot be measured directly, since the result of analyses at various times will give a series of values of concentration (x) corresponding to various times (t). There are, however, two methods that might be employed to derive the rate constant of the reaction.

The first method is to plot a graph of x against t and measure the slope of the curve at various values of x. This gives dx/dt (see M18) which can then be substituted directly into the rate law to give the rate constant. For example, if the rate law is $dx/dt = k(a - x)$ then

$$k = \frac{dx/dt}{(a - x)} \qquad (2.1)$$

This method of calculation is used in Section 13.1.

The objection to the general use of this method, which is mathematically simple, is that it lacks in accuracy because of the difficulty of measuring the slope of a curve. This difficulty is accentuated if the curve is not completely smooth owing to experimental errors in some of the points.

The numerical measurement of the slope of the curve is avoided, in the second method, by integrating the rate equation before the experimental results are substituted into it. This converts the rate law from the differential form to an equation of the type

$$kt = \text{function of } x = F(x), \text{ say} \qquad (2.2)$$

Using the function $F(x)$ derived below for the appropriate order of reaction, k can be found directly by substituting values of x and t into equation 2.2. Small experimental errors can be dealt with by plotting a graph of $F(x)$ against t and (either visually, or by the method of least squares) drawing the best straight line through the origin and the points. Using M3, the slope of the line is equal to k.

This chapter deals only with the integration of the simpler types of rate law and therefore omits consideration of reversible reactions and reactions which occur in more than one stage.

2.2 Zero-order Reactions

These reactions are not common, but they do occur in some heterogeneous systems (page 102) and in some solutions (page 76). The appropriate rate law for a zero-order reaction is

$$\text{rate} = \text{constant}$$

or, in mathematical symbols

$$\frac{dx}{dt} = k \tag{2.3}$$

where x is the concentration of product formed. Integrating with respect to t (using M20)

$$x = kt + \text{constant}$$

Since $x = 0$ when $t = 0$ (i.e. at the beginning of the reaction, no product has formed) the constant must be zero. Hence

$$x = kt \tag{2.4}$$

The dimensions of k are concentration/time, i.e. mol 1^{-1} s^{-1}.

To calculate the half-life of this reaction the condition that $t = t_{1/2}$ when $x = a/2$ (a being the initial concentration of the reactant) is substituted into equation 2.4 to give

$$a/2 = kt_{1/2} \tag{2.5}$$

Thus in a zero-order reaction the half-life is proportional to the initial concentration of the reactant.

2.3 First-order Reactions

These reactions are common and are often observed in solution where the solvent happens to be one of the reactants. Many gas-phase reactions and radioactive decay processes also obey the first-order rate law

$$\frac{dx}{dt} = k(a - x) \tag{2.6}$$

Rearrangement gives

$$\frac{dx}{(a - x)} = k\, dt \tag{2.7}$$

Integrating, using M21

$$-\ln(a - x) = kt + \text{constant} \tag{2.8}$$

When $t = 0$, $x = 0$ and hence $-\ln a = \text{constant}$. Substituting for the constant in equation 2.8

$$\ln a - \ln(a - x) = kt \tag{2.9}$$

or, using M14

$$
\begin{aligned}
kt &= \ln \frac{a}{a - x} \\
\text{or} \\
kt &= \ln(n_0/n)
\end{aligned}
\tag{2.10}
$$

where n_0 is the initial number of atoms or molecules of the reactant and n is the number remaining at time t. It will be seen from equation 2.10 that the value of k depends only on the *ratio* of two concentrations, the dimensions of k being reciprocal time, e.g. s^{-1}. The important practical point that follows from this is that the first-order rate constant can be calculated from the *relative values of the concentrations* at various times (see M16). Thus the first-order reaction

$$RBr + C_2H_5OH \rightarrow ROC_2H_5 + HBr$$

(where R is an acyl group) can be followed by removing aliquots, cooling them rapidly and titrating the liberated HBr with alkali. The concentration of the alkali is not needed in order to find k, since the ratio $a/(a - x)$ which is required in equation 2.10 can be expressed as the ratio of two titrations. If X_∞ is the titration after 'infinite' time and X is the titration after time t then

$$\frac{a}{a - x} = \frac{X_\infty}{X_\infty - X}$$

The use of this type of calculation is shown in Section 13.3.

First-order reactions have another important property which is that the half-life (i.e. the time for 50% reaction) does not depend on the initial concentration. This may be seen by substituting in equation 2.10 the conditions for the half-life, namely that when $t = t_{1/2}$, $x = a/2$, whence

$$\ln 2 = kt_{1/2}$$

i.e.

$$t_{1/2} = \frac{\ln 2}{k} = \frac{0.6932}{k} \qquad (2.11)$$

The half-life is thus independent of the initial concentration of the reactants. In other words, provided a reaction obeys strictly the first-order equation, the time for 50% reaction cannot be changed by altering the initial concentrations. For batch processes this means that the batch time cannot be decreased by increasing the concentration of the reactants. For first-order reactions the half-life is directly related to the rate constant by an expression (2.11) that is independent of the initial concentration.

The half-life is an easier concept to visualize than a rate constant, and so in the case of radioactive decay, where the processes are all first order, rates are expressed in terms of half-lives. Thus the statement that the half-life of radium is 1690 years means that its first-order rate constant k is 1.301×10^{-11} s^{-1}.

Unfortunately, in all chemical reactions with order other than unity the half-life depends on the initial concentration. Thus to be consistent it is preferable to use rate constants for all orders of reaction.

The rate constant of a first-order reaction can be shown to be equal to the reciprocal of the average life of the reacting molecules or atoms.

2.4 Pseudo-first-order Reactions

These are bimolecular reactions in which one component is present in large excess, e.g. if the solvent is one of the reactants. Although the reaction is bimolecular, the experimental results will obey a first-order law as it is impossible to detect the change in concentration brought about by the reaction of the component in excess. For example, the molarity of water is $1000/18 = 55.6$. If an 0.1 molar solution of ethyl acetate is completely hydrolysed according to equation (a) below, the water molarity will be reduced by 0.1 in 55.6, i.e. by less than 0.2%.

The fact that a reaction of this type is bimolecular can often be shown by adding an inert solvent. Thus in reaction (b), if ethyl alcohol is replaced by a dilute solution of ethyl alcohol in benzene, the reaction rate will depend on the ethyl alcohol concentration, thus giving a second-order rate law.

Two examples of pseudo-first-order reactions are therefore

(a) $\quad CH_3CO_2C_2H_5 + H_2O \rightarrow CH_3CO_2H + C_2H_5OH$

(b) $\quad CH_3COCl + C_2H_5OH \rightarrow CH_3CO_2C_2H_5 + HCl$

2.5 Second-order Reactions

This is the most common order of reaction and in general it may be said that, provided a reaction occurs in a single step and the reactants are present in roughly equal concentrations, a second-order reaction may be expected. Reactions of this type occur often in organic chemistry. A simple second-order reaction is

$$A + B \rightarrow products$$

in which one molecule of A reacts with one molecule of B. If x denotes the decrease in concentration of A and B at time t, and *a* and *b* are the initial concentrations, the rate law states that the rate is proportional to the concentration of both reactants, i.e.

$$\frac{dx}{dt} = k(a - x)(b - x) \qquad (2.12)$$

First consider the special case in which the initial concentrations of A and B are equal. Equation 2.12 then becomes

$$\frac{dx}{dt} = k(a - x)^2 \qquad (2.13)$$

Rearranging

$$\frac{dx}{(a - x)^2} = k \, dt$$

Integrating using M20

$$\frac{1}{(a - x)} = kt + \text{constant} \qquad (2.14)$$

$x = 0$ when $t = 0$, whence $1/a = $ constant. Substituting for the constant in 2.14

$$kt = \frac{1}{(a - x)} - \frac{1}{a}$$

i.e.

$$kt = \frac{x}{a(a - x)} \qquad (2.15)$$

To evaluate graphically the rate constant of a second-order reaction (with equal concentrations of reactants) a graph is plotted of $x/(a - x)$ against time, when a straight line through the origin is obtained with slope equal to ak. As in the case of first-order reactions, the fraction $x/(a - x)$, which is required for the graph, may be expressed as a ratio of concentrations, but the particular value of the initial concentration is required to convert the slope of the graph into the rate constant. An example of this type of calculation is shown in Section 13.3.

The half-life can be found by substituting the usual condition that $t = t_{1/2}$ when $x = a/2$ into the integrated rate equation 2.15. This gives

$$kt_{1/2} = \frac{1}{a}$$

i.e.

$$t_{1/2} = \frac{1}{ka} \tag{2.16}$$

Thus, in contrast to the result for first-order reactions (2.11), the half-life of a second-order reaction is inversely proportional to the initial concentration of reactants, i.e. increasing the concentration will decrease the time taken to reach 50% reaction.

In dealing with second-order reactions in which the initial concentrations are *not equal*, the half-life must be expressed relative to one of the reactants, and it is not possible to derive a simple expression for half-life in terms of initial concentration and rate constant. Nevertheless, it is still true to say that an increase in initial concentration will decrease the time for a given amount of reaction.

The general form of the second-order rate law is

$$\frac{dx}{dt} = k(a - x)(b - x) \tag{2.17}$$

which may be integrated as follows. Rearranging equation 2.17

$$\frac{dx}{(a - x)(b - x)} = k \, dt \tag{2.18}$$

Rearranging 2.18, using partial fractions, gives

$$\frac{1}{a - b} \left[\frac{1}{b - x} - \frac{1}{a - x} \right] dx = k \, dt \tag{2.19}$$

which on integrating (using M21) gives

$$(a - b)kt = \ln(a - x) - \ln(b - x) + \text{constant} \tag{2.20}$$

Using the condition that $x = 0$ when $t = 0$, and M14

$$\text{constant} = \ln \frac{b}{a} \tag{2.21}$$

Substituting 2.21 in 2.20 and rearranging

$$(a - b)kt = \ln \frac{b(a - x)}{a(b - x)} \tag{2.22}$$

k may be evaluated from 2.22 by plotting a graph of $\log[(a - x)/(b - x)]$ against time, when a straight line passing through the origin with slope equal to $(a - b)k/2.303$ is obtained. A numerical example of the use of equation 2.22 is given in Section 13.4.

Whenever the initial concentrations of a and b are not equal, equation 2.22 must be used to calculate the value of k.

It sometimes happens in a second-order reaction that more than one molecule of one reactant (A) is consumed for each molecule of B, as in the reaction

$$2A + B \rightarrow \text{products}$$

In this case, if x represents the change in concentration of B the rate law becomes

$$\frac{dx}{dt} = k(a - 2x)(b - x) \tag{2.23}$$

the integrated form of which is

$$(a - 2b)kt = \ln \frac{b(a - 2x)}{a(b - x)} \tag{2.24}$$

A bimolecular second-order reaction that obeys this rate law is the reaction of 2,4-dinitrochlorobenzene with aniline in benzene. Denoting the first compound as RCl the reaction occurs as follows:

$$RCl + C_6H_5NH_2 \rightarrow C_6H_5NHR + HCl$$

$$HCl + C_6H_5NH_2 \rightarrow C_6H_5NH_3Cl \text{ (aniline hydrochloride)}$$

The aniline hydrochloride does not react with the 2,4-dinitrochlorobenzene, and so for each molecule of the latter that reacts two molecules of aniline are removed. Thus the modified rate law 2.24 is obeyed.

2.6 Third-order Reactions

Reactions of this order are uncommon, some well-established cases being concerned with the reactions of nitric oxide. Termolecular reactions are theoretically expected to be uncommon since the probability of a collision involving three molecules with sufficient energy to react and with correct orientations is very slight.

In the simple case of a third-order reaction in which all the reactants are at the same initial concentration a, the rate law becomes

$$\frac{\mathrm{d}x}{\mathrm{d}t} = k(a - x)^3 \qquad (2.25)$$

Rearranging

$$\frac{\mathrm{d}x}{(a - x)^3} = k \, \mathrm{d}t \qquad (2.26)$$

Integrating 2.26 using M21

$$\frac{1}{2(a - x)^2} = kt + \text{constant} \qquad (2.27)$$

At the beginning of the reaction $x = 0$ when $t = 0$. Thus

$$\text{constant} = \frac{1}{2a^2}$$

Hence from equation 2.27

$$kt = \frac{1}{2(a - x)^2} - \frac{1}{2a^2} \qquad (2.28)$$

The half-life can be found by substituting the condition

$$x = \frac{a}{2} \text{ when } t = t_{\frac{1}{2}}$$

in equation 2.28. This gives

$$kt_{\frac{1}{2}} = \frac{3}{2a^2} \qquad (2.29)$$

The half-life is therefore inversely proportional to the square of the initial concentration.

There are no known reactions whose molecularity is four, but there are a few fourth-order reactions in solution. One of the best known is the reaction of bromide and bromate ions in acid solution, which is discussed in Section 7.7.

Comparison of equations 2.5, 2.11, 2.16 and 2.29 shows that the dependence of half-life on concentration varies in a regular

Table 2.1

Order	Rate law in differential form	Rate law in integrated form	Dimensions of k	Half-life proportional to
0	$\dfrac{dx}{dt} = k$	$kt = x$	mol l^{-1} s^{-1}	a^1
1	$\dfrac{dx}{dt} = k(a - x)$	$kt = \ln \dfrac{a}{(a - x)}$	s^{-1}	$a^0 \, (= 1)$
2	$\dfrac{dx}{dt} = k(a - x)^2$	$kt = \dfrac{x}{a(a - x)}$	$l \, mol^{-1} \, s^{-1}$	a^{-1}
3	$\dfrac{dx}{dt} = k(a - x)^3$	$kt = \dfrac{1}{2(a - x)^2} - \dfrac{1}{2a^2}$	$l^2 \, mol^{-2} \, s^{-1}$	a^{-2}
2	$\dfrac{dx}{dt} = k(a - x)(b - x)$	$kt = \dfrac{1}{a - b} \ln \dfrac{b(a - x)}{a(b - x)}$	$l \, mol^{-1} \, s^{-1}$	—

way as the order is changed. This is shown, together with a summary of the main points in this chapter, in *Table 2.1*.

2.7 Determination of Order of Reaction

The two most convenient methods of determining the order of a reaction from the experimental results are the method of empirical fit and the half-life method.

2.7.1 THE METHOD OF EMPIRICAL FIT

It is assumed that experimental data are available in the form of a series of values of x at different times t, including the initial value when $t = 0$. These data are all substituted into each of the rate laws in turn until a law is found that gives constant values of k, or that gives a straight line plot when the appropriate function of x is plotted against time. Thus, if it is suspected that a reaction is of the first order, a graph of $\log(a - x)$ against time is plotted. If the result is a straight line the reaction is in fact first order. If the results do not, however, fit the first-order law, they must be substituted into the other rate equations until one is found that gives a constant value of k.

This procedure is illustrated in *Figure 2.1*, in which data, calculated to fit a second-order rate law, have been plotted in turn according to the zero-, first- and second-order laws. The initial concentrations of the two reactants are 0.100 mol 1^{-1}. The value of k from the slope of the straight line graph (using equation 2.15) is 2.18×10^{-3} 1 mol^{-1} s^{-1}.

It will be seen that over a small fraction of the reaction the plots for the various orders are all nearly linear. Hence to establish the order by this method it is essential to make an experimental study over as large a fraction of the reaction as possible.

2.7.2 THE HALF-LIFE METHOD

This method involves measuring the rate of a reaction several times at the same temperature, varying the initial concentration of the reactants. To simplify the mathematical treatment, all reactants used are made up initially at the same concentration so that one of the first four rate laws in *Table 2.1* applies. From the experimental results, a graph of percentage reaction against time is plotted, and by interpolation the times for 50% reaction (half-life)

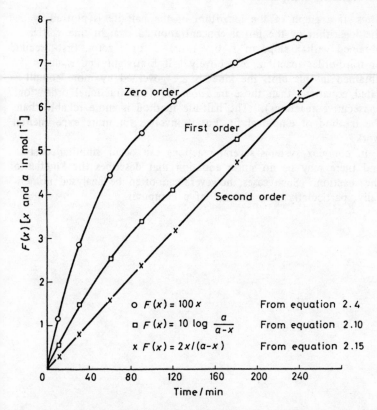

Figure 2.1 Determination of order of reaction. Plot of F(x) against time for zero-, first- and second-order rate laws

are found. By combining the results in *Table 2.1*, it is seen that if n equals the order of reaction, then the half-life of a particular order of reaction is given by

$$t_{1/2} \propto a^{1-n}$$

(using M1)

$$t_{1/2} = Ca^{1-n} \tag{2.30}$$

where C is a proportionality constant.

Taking logarithms of 2.30 (using M13)

$$\log t_{1/2} = \log C + (1 - n) \log a \tag{2.31}$$

Thus, if a graph of the logarithm of the half-life is plotted against the logarithm of the initial concentration, a straight line will be obtained with a slope of 1, 0, -1 or -2 for a zero-, first-, second- or third-order reaction, respectively. If a straight line is not obtained in this plot, the kinetics are governed by more complicated equations than those in *Table 1.1* (e.g. fractional orders or consecutive reactions). The half-life method is more reliable than the method of empirical fit, but it involves far more experimental work.

In complex systems several reactions can occur simultaneously, and there may be no single equation that describes the kinetics of the reaction. Such cases, however, can often be analysed numerically, particularly with the aid of a computer.

3

BASIC EXPERIMENTAL METHODS

This chapter is concerned with the standard experimental techniques normally available in a teaching laboratory for the measurement of reaction rates. The specialized techniques of 'fast reactions' are discussed separately in Chapter 12. The numerical analysis of the experimental results obtained from kinetic studies is discussed in Chapter 13.

3.1 General Points

In essence the experimental problem in reaction kinetics is to devize an analytical technique which will enable the concentrations of the reactants to be found at any time during the reaction. Full details of the laboratory procedures may be found in textbooks of practical physical chemistry. An attempt will be made here to outline some of the principal features involved.

The rates of chemical reactions vary markedly with temperature and as a rough working rule a 5–10% increase in rate per degree centigrade rise in temperature may be assumed. To obtain rate constants to an accuracy of 2% the temperature must be held constant to within 0.1 °C. Hence the first requirement for kinetic studies is a bath fitted with an accurate means of controlling the temperature. Many commercial control units are available that are suitable for kinetic work. The temperature of the bath should be measured with a thermometer that has been compared with a National Physical Laboratories standard.

An appreciable amount of reaction can occur in the time taken for reactants to reach bath temperature from room temperature, so the experiment must be started with particular care. The usual procedure is to warm separately solutions of the two reactants until they have reached the bath temperature. They are then rapidly mixed and stirred and an aliquot is removed immediately before the clock is started. The time at which the clock is started is known as 'zero time'. The concentrations of the reactants at 'zero time' can then be calculated from the titration of the first aliquot removed. This technique makes allowance for the small amount of reaction that occurs in the mixing process.

A further complication in reactions at high temperatures is due to the expansion of the solvent, which lowers the concentration of the solution. For accurate work this expansion must be allowed for (see Section 13.4).

When using a sampling technique it must be possible to 'freeze' (i.e. to stop suddenly) the reaction in the sample so that in the time taken to carry out the analysis no further product is formed. For reactions at high temperatures this may be done by rapid cooling in iced water, but for reactions occurring near room temperature one of the reactants must be removed chemically. Thus an acid-catalysed reaction may be 'frozen' by running the sample into a solution containing excess alkali.

The initial concentrations of the reactants are most accurately determined by weighing a known amount of substance directly into a volumetric flask and making up to the mark with the appropriate solvent. Alternatively, a solution of approximately known concentration may be prepared, and an aliquot of this removed and analysed by one of the usual volumetric techniques.

A laboratory stop-clock is sufficiently accurate for measuring the time of reaction, since an error as large as a quarter of an hour per day is equivalent to only a 1% error in the time measurement.

Several common methods of analysis used in kinetic studies are listed below, together with some comments on their applicability.

3.2 Sampling Methods

There are two main ways of removing samples at known times. In the first, a large volume of the reaction mixture is made up initially and placed in the temperature-controlled bath. At various times aliquots are withdrawn (e.g. with a pipette previously warmed to the bath temperature). The sample removed is 'frozen' and then analysed, usually by a volumetric method. This sampling method is most frequently used for reactions near room temperature.

For reactions at high temperatures, the vapour pressures of the solvents become appreciable, and to prevent solvent loss, and also for general convenience, the 'sealed-tube method' is used. In this a number of aliquots of the reactant mixture, at room temperature, are placed into glass tubes and sealed. These tubes are then all immersed in the bath at the same time. After a few minutes, when the tubes have reached bath temperature, one is removed and 'frozen' by cooling and the stop-clock is started. Analysis of the first tube enables the initial concentration to be corrected for the reaction before zero time. The remaining tubes are then removed at suitable times, and analysed. Since the sealed-tube technique is used for reactions at high temperatures, simple cooling of the tube in iced water will decrease the rate sufficiently to prevent further reaction taking place while the analysis is carried out. A variation of the sealed-tube method is described in Section 13.3.

If a reaction at room temperature is followed by the sealed-tube method, it can be 'frozen' by placing the tubes in a mixture of acetone and solid carbon dioxide at $-78\,^{\circ}C$. If this is done the tubes can be left for many hours before analysis, with no significant reaction on storage.

A reaction that can be conveniently followed by the first method is the hydrolysis of ethyl acetate by sodium hydroxide in water at $25\,^{\circ}C$

$$CH_3CO_2C_2H_5 + NaOH \rightarrow CH_3CO_2Na + C_2H_5OH$$

The reaction is followed by removing aliquots and adding them to a known excess of dilute hydrochloric acid, which stops the reaction. The excess of acid is then back-titrated with sodium hydroxide.

An appropriate reaction for the sealed-tube method is the reaction of sodium ethoxide with n-butyl bromide in ethanol at $60\,^{\circ}C$

$$n\text{-}C_4H_9Br + NaOC_2H_5 \rightarrow NaBr + n\text{-}C_4H_5OC_2H_5$$

Here the tubes can be rapidly cooled in a freezing-mixture and then broken under dilute nitric acid. The resultant bromide ions are titrated with standard silver nitrate solution, thus giving the amount of product formed.

Sometimes a sample may be removed and analysed by measuring one of its physical properties. Thus the acid hydrolysis of sucrose to form fructose and glucose can be followed by determining the refractive index of successive samples. The kinetics of polymerization is often followed by measuring the increase in viscosity as reaction proceeds. In both cases, however, calibration is necessary to relate the change in the physical property to percentage reaction, since the measured properties are not linearly related to concentration.

3.3 Continuous Methods

These methods are based on the measurement of a physical property during the course of a reaction without removing a sample or disturbing the reaction mixture. The following are some of the techniques that have been used in kinetic studies:

1. solution conductance
2. optical rotation
3. measurement of pressure or volume of gas evolved
4. dilatometry
5. spectrophotometry
6. spectrofluorimetry
7. potentiometry

Before using a continuous method, independent chemical analyses should be made to ensure that there is proportionality between the property being measured and the concentration of the reactant. If this is not found to be so, a calibration curve must be drawn which relates the property to be measured to percentage reaction. This curve is then used in interpreting the experimental results.

3.3.1 SOLUTION CONDUCTANCE

This is applicable to any reaction involving either an increase or a decrease in the number of ions, or the replacement of one ion by another with a different ionic conductance. If the solutions are sufficiently dilute it may be assumed that the change in conductance is proportional to the percentage reaction. A numerical example illustrating this is shown in Section 13.2. In more concentrated solutions a calibration curve must be prepared.

It may be noted here that the ionic conductances of the hydrogen and the hydroxide ions are much larger than the conductances of other ions. Hence any reaction producing an increase or decrease in the number of these ions will be capable of being followed conductometrically.

Some examples of reactions where this type of technique is applicable are

$$CH_3CO_2C_2H_5 + NaOH \rightarrow C_2H_5OH + CH_3CO_2Na \qquad (3.1)$$

$$t\text{-}C_5H_{11}I + H_2O \rightarrow t\text{-}C_5H_{11}OH + HI \qquad (3.2)$$

In equation 3.1 the conductance falls with time as the highly conducting hydroxide ions are converted into acetate ions of much lower conductance. In equation 3.2 ions are formed in the reaction and so the conductance increases. In both cases it is necessary to allow the reaction to proceed for 'infinite time' (see

Section 1.9) so that the contribution to the conductance due to the product may be found.

If G = conductance at time t,
G_0 = initial conductance and
G_∞ = final conductance, at 'infinite time',

then assuming that change in conductance is proportional to percentage reaction, the amount of reaction at time t is proportional to $G - G_0$. The total amount of reaction is proportional to $G_\infty - G_0$. Therefore

$$\text{fraction of reaction time at time } t = \frac{G - G_0}{G_\infty - G_0} = \frac{x}{a} \qquad (3.3)$$

For a first-order reaction, where the rate constant depends only on a ratio of concentrations, equation 3.3 is sufficient to determine the rate constant, as the value of x/a may be substituted into equation 2.10. A graphical method, as in Section 13.2, can also be used. For all other orders the initial concentrations of reactants must be obtained.

3.3.2 OPTICAL ROTATION

In this method the angle through which plane-polarized light is rotated by the reaction mixture is measured. It is therefore limited to those reactions involving optically active substances, and has been used extensively in the hydrolysis of sucrose to give glucose and fructose.

The theory of the calculations from optical rotation measurements is the same as that from conductance given above and so numerical calculations are similar to Section 13.2:

let X = rotation at time t
X_0 = initial rotation
X_∞ = final rotation, then as in the case of the conductance measurements,

$$\text{fraction of reaction at time } t = \frac{X - X_0}{X_\infty - X_0} = \frac{x}{a}$$

3.3.3 REACTIONS INVOLVING GASES

If a gas is liberated in a reaction, the amount of reaction may be conveniently followed by measuring either the volume of gas formed at constant pressure, or the pressure produced by the gas at constant volume. Some examples of reactions which have been followed by this method are

$$C_6H_5N_2Cl + H_2O \xrightarrow{\text{acid}} N_2 + C_6H_5OH + HCl$$

$$CO(CH_2CO_2H)_2 \xrightarrow{\text{acid}} CO(CH_3)_2 + 2CO_2$$

$$C_2H_5NH_2 \xrightarrow{\text{heat}} C_2H_4 + NH_3$$

It is usual to work at moderately low pressures (up to one atmosphere), and for easily liquefiable gases, high temperatures must be used. Under these conditions the ideal gas law can be applied, and the number of moles (n) involved may be calculated using the equation

$$n = pV/RT$$

where p, V, T are the pressure, volume and absolute temperature of the gas, and R is the gas constant.

A typical calculation is given in Section 13.1.

3.3.4 DILATOMETRY

In dilatometry the change in volume produced by reaction is measured by placing the reaction mixture in a completely filled reaction vessel connected to a capillary tube. The level of liquid in the capillary is altered if the volume changes. This method is best suited to concentrated solutions where volume changes are greatest. As the dilatometer has effectively the same construction as a thermometer, temperature control to ±0.001 °C is essential. The method has been applied to the hydration of isobutene at 25 °C

$$(CH_3)_2C=CH_2 + H_2O \xrightarrow{\text{acid}} (CH_3)_3COH$$

in which there is a small decrease in volume.

It is assumed that the percentage reaction is proportional to movement of the liquid in the capillary.

3.3.5 SPECTROPHOTOMETRY

Spectrophotometry refers to the measurement of the intensity of light transmitted by a solution at various wavelengths. If the products of a chemical reaction absorb to a greater or lesser extent than the reactants, then the reaction can be followed spectrophotometrically. An example of such a reaction is the alkaline hydrolysis of methyl salicylate

| maximum absorbance | maximum absorbance |
| at 332 nm | at 305 nm |

The basic form of the methyl salicylate absorbs strongly in the ultraviolet region of the spectrum, but at a higher wavelength than the salicylate ion. At the wavelengths involved (332 nm for the basic methyl salicylate) the methanol has negligible absorbance. The reaction can, therefore, be followed by measuring the decrease in absorbance at 332 nm. However, at this wavelength the absorbance of the salicylate ion is not negligible. The system must, therefore, be analysed spectrophotometrically as a two-component mixture.

3.3.6 SPECTROFLUORIMETRY

The instrumentation of spectrofluorimetry is similar to that used in spectrophotometry except for two principal modifications. Firstly, the detector is placed at an angle perpendicular to the direction of the irradiating beam. Secondly, an additional monochromator is placed before the detector. Thus all radiation, except that being emitted and monitored from the fluorescing substance, is prevented from reaching the detector.

Since fluorescent emissions are generally very intense, fluorimetric analysis can, in some instances, detect substances at concentrations as low as one part in 10^{10}. This is a sensitivity approximately 1000 times greater than that obtainable by absorbance spectrophotometry.

For example, in enzyme reactions it is often necessary to monitor reaction progress at concentrations too low for absorbance spectroscopy to be applicable. Because many co-enzymes fluoresce naturally, or can be converted into fluorescent compounds by simple chemical reactions, spectrofluorimetry ideally suits the monitoring requirements.

In the conversion of ethanol to ethanal, catalysed by alcohol dehydrogenase, the oxidant is the co-enzyme NAD (nicotinamide adenine dinucleotide). The oxidized form NAD^+ is non-fluorescent, but the emergent reduced form NADH fluoresces strongly. The emission is monitored at 436 nm.

$$C_2H_5OH + NAD^+ \rightarrow CH_3CHO + NADH + H^+$$

Various kinetic techniques use fluorescent emission. Some, for example, measure the quenching effect (p. 148) which a reactant may have on a compound that fluoresces.

3.3.7 POTENTIOMETRY

Potentiometric methods have been applied to the study of fast reactions, such as the bromination of phenols. The potentiometric method measures the e.m.f. of a cell, which depends on the *logarithm* of the concentration rather than the concentration itself. This can be an advantage in first-order reactions, where the logarithm of the amount of reactant remaining varies linearly with time. This can be seen from equation 2.9.

It may be helpful to recall that in dilute solutions the potential of a hydrogen electrode varies linearly with the logarithm of the hydrogen ion concentration and therefore with the pH value of the solution. This is the basis of electrochemical methods of pH measurement.

In the potentiometric investigation of the kinetics of the bromination of phenols, a bright platinum indicator electrode and a saturated calomel reference electrode are immersed in the reaction mixture. In addition to bromine and phenol, this solution also contains a large known excess of bromide ions, so that the measured cell e.m.f. depends only on the concentration of bromine remaining in the solution. The theory of this method and the calculation of the rate constant from observed e.m.f. changes are shown in Section 13.6.

THE EFFECT OF TEMPERATURE ON REACTION RATES

4.1 The Nature of the Problem

There is one fundamental problem in reaction kinetics that has not yet been dealt with. From his very first studies the chemist is brought up to believe that all molecules of a given substance are identical (when the existence of isotopes is ignored). Therefore, it would be expected that all molecules of this substance should behave in the same way in a chemical reaction. But experiment shows that this cannot be so. Chemical reactions occur at a *finite* rate, and whereas one molecule may react immediately, another of the same type may have to wait several hours before it can react. In short, 50% of the reacting molecules *must* wait for more than the half-life before their turn for reaction occurs.

If it is accepted that all molecules of the same substance are equivalent, it might be expected that either no reaction would occur or that every collision between reactant molecules would lead to reaction. In the latter case, calculations show that the frequency of collisions is so high that all reactions would be virtually instantaneous (as in fact is the case for the reaction of hydrogen ions with hydroxide ions in aqueous solutions). From this viewpoint it is difficult to see how intermediate rates of reaction can occur.

Arrhenius solved this problem in 1897 by postulating that normal chemical molecules do not take part in chemical reactions. Only those molecules which have acquired more than a certain critical energy, called the *energy of activation*, are able to react. The activated molecules are extremely few in number and arise as a result of random thermal collisions between molecules which occasionally give a molecule many times the average energy.

It may be helpful at this point to draw an analogy between the distribution of energy in a molecular system and the distribution of money among men. In both cases the rich are few and the poor are many! If one imagines that the average income is, say, £5000 a year, it will be easily realized that the fraction of the population with an income of more than ten times the average will be relatively small, while the fraction with an income of one hundred times the average (£500 000) will be extremely small but *not* zero.

Accepting this analogy, it will be seen that the activated molecules which possess about twenty times the average energy will be present in very small numbers, so an activated molecule is a rare phenomenon. The chance of a molecule becoming activated by a single collision is usually much less than the chance that the first stranger that the reader meets will be a millionaire.

It may now be seen why it is possible for chemical reactions to occur slowly. Even though there may be thousands of millions of collisions per molecule per second, the chance of a molecule becoming an 'energy millionaire' (i.e. activated) is so remote that it may take several minutes, hours or even days before the molecule can react.

This analogy may be taken a stage further. Physiologically, the average man is not significantly different from the £500 000 a year man, but the economic influence of the rich man is much greater than that of the poor man. Similarly, an activated molecule does not differ from an average molecule if the energy of activation is removed, but while the molecule possesses this activation energy its chemical reactivity is completely different from that of an average molecule. Perhaps one can paraphrase the famous quotation from George Orwell's *'Animal Farm'* to make it say: 'All molecules are equal, but some are more equal than others'!

4.2 The Concept of Activation Energy

The idea that the rate of certain processes is controlled by the ability to overcome a critical energy barrier, i.e. an activation energy, has become a major theoretical concept in science. This concept has been applied to processes as varied as radioactive decay, viscous flow and the resistance of transistors.

Using the idea of activation, Arrhenius derived an equation to describe the variation of rate constant with temperature. This was another puzzling feature of reaction kinetics, for the rates of most reactions at room temperature increase by a factor of two to three for a 10 °C temperature rise, whereas the kinetic energy of a gas increases by only 3% for the same temperature rise. This shows that the reaction rate increases much more rapidly than the average energy and hence the average energy cannot be the factor that is determining the rate of reaction.

The original derivation of Arrhenius's equation is as follows. Consider a reversible second-order reaction such as

$$A + B \underset{k_{-1}}{\overset{k_1}{\rightleftharpoons}} C + D \tag{4.1}$$

Having attained dynamic equilibrium, then

rate of forward reaction = rate of back reaction

From the Law of Mass Action these rates are proportional to the products of the concentrations of the reactants, i.e. using M1

$$k_1 ab = k_{-1} cd$$

where a, b, c, d represent the concentrations of A, B, C, D, and k_1 and k_{-1} are proportionality constants. Rearranging this equation

$$\boxed{\frac{cd}{ab} = \frac{k_1}{k_{-1}} = K} \tag{4.2}$$

since the ratio of the two constants is also a constant (K). The above equation is a mathematical expression for a *classical equilibrium constant*. The word 'classical' implies that the equilibrium constant is calculated in terms of concentrations. This is in contrast to the *thermodynamic equilibrium constant*, which is calculated in terms of activities.

It is imagined that when two molecules, which between them possess the activation energy, undergo a collision, they pass into a 'transition' state which is intermediate between the reactants and the products. In *Figure 4.1* the energies E of the reactants, products and transition state are shown diagrammatically. The distance

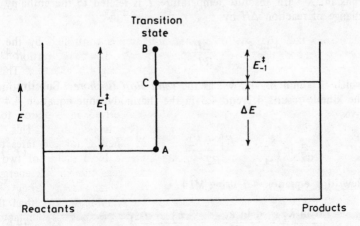

Figure 4.1 Energy diagram

AB corresponds to the difference in energy between the transition state and the reactants, which is the activation energy of the forward reaction $E_1^{\ddagger}$ Similarly, BC corresponds to the activation energy of the back reaction, $E_{-1}^{\ddagger}$. So, by geometrical construction

$$AB - BC = AC$$

The net energy change of reaction is

$$\Delta E = E_1^{\ddagger} - E_{-1}^{\ddagger}$$

If there is negligible change in volume on reaction, such as for reactions in solution under conditions of constant temperature and pressure

$$\Delta E = \Delta H$$

where ΔH is the enthalpy change of the reaction. Under such conditions therefore

$$\Delta H = E_1^{\ddagger} - E_{-1}^{\ddagger} \tag{4.3}$$

4.3 The Arrhenius Equation

It was shown by van't Hoff that the variation of the equilibrium constant K with absolute temperature T is related to the enthalpy change of reaction ΔH by

$$\frac{\mathrm{d} \ln K}{\mathrm{d}T} = \frac{\Delta H}{RT^2} \tag{4.4}$$

which is commonly known as the *van't Hoff isochore*. Substituting the kinetic results 4.2 and 4.3 in the thermodynamic equation 4.4 gives

$$\frac{\mathrm{d}}{\mathrm{d}T} \ln \frac{k_1}{k_{-1}} = \frac{E_1^{\ddagger} - E_{-1}^{\ddagger}}{RT^2} \tag{4.5}$$

Rewriting equation 4.5 using M14

$$\frac{\mathrm{d} \ln k_1}{\mathrm{d}T} - \frac{\mathrm{d} \ln k_{-1}}{\mathrm{d}T} = \frac{E_1^{\ddagger}}{RT^2} - \frac{E_{-1}^{\ddagger}}{RT^2} \tag{4.6}$$

An equation of this sort suggests that the forward and back reactions have independent kinetic effects and thus Arrhenius suggested that the equation might be split to give

$$\frac{d \ln k_1}{dT} = \frac{E_1^{\ddagger}}{RT^2} + I \quad \text{and} \quad \frac{d \ln k_{-1}}{dT} = \frac{E_{-1}^{\ddagger}}{RT^2} + I \qquad (4.7)$$

where I is a constant.

Arrhenius found by experiment that for a given reaction the value of the constant I was zero. The variation of the rate constant k with temperature can be expressed satisfactorily by the simplified equation

$$\boxed{\frac{d \ln k}{dT} = \frac{E^{\ddagger}}{RT^2}} \qquad (4.8)$$

Equation 4.8 is one form of the Arrhenius equation. It may be integrated using M21 to give

$$\boxed{\ln k = -E^{\ddagger}/RT + \text{constant}} \qquad (4.9)$$

assuming that $E^{\ddagger}$ does not vary with temperature. Or, using M15

$$\boxed{k = A \exp\left(\frac{-E^{\ddagger}}{RT}\right)} \qquad (4.10)$$

where A is a constant, known as the *pre-exponential factor* or the *frequency factor*. It is sometimes referred to simply as the A-factor. A has the same units as the rate constant, and would be numerically equal to the rate constant in a reaction with zero energy of activation (see M6).

Equations 4.8, 4.9 and 4.10 are three equivalent ways of expressing the same result. It is important that the reader should be able to recognize the Arrhenius equation in any of these three mathematical forms.

4.4 Explanation of the Effect of Temperature

The activation energy represents the energy that the reactant molecules must acquire before they can form the transition state. This quantity cannot be measured *directly*. However, an examination of equation 4.9 shows that $E^{\ddagger}$ may be determined if the rate constant, or another parameter proportional to it (see M16), is known at several temperatures. For, using M3, a plot of ln k against the reciprocal of the absolute temperature $1/T$ gives a straight line of slope $-E^{\ddagger}/R$. Alternatively, if common logarithms (to the base ten) are used, a plot of log k against $1/T$ will still be a straight line, but using M17 the slope will be $-E^{\ddagger}/2.303R$. Operationally, therefore, the activation energy of any one reaction is *defined* in terms of the Arrhenius equation. The quantity so measured is often called the *Arrhenius activation energy*.

Rate constants of many chemical reactions have been measured over a range of temperatures. In the vast majority of cases the plot of log k against $1/T$ does in fact give a straight line. This is illustrated in *Figure 4.2* by the plot of data for the decomposition of acetonedicarboxylic acid in aqueous solution

$$HO_2CCH_2COCH_2CO_2H \rightarrow CH_3COCH_3 + 2CO_2$$

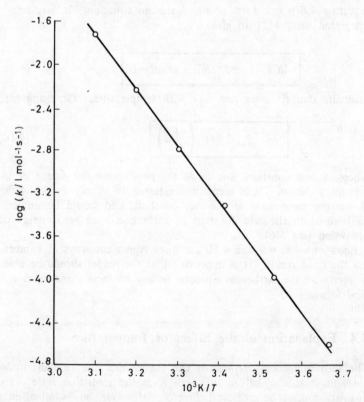

Figure 4.2 Decomposition of acetonedicarboxylic acid in aqueous solution (Arrhenius plot)

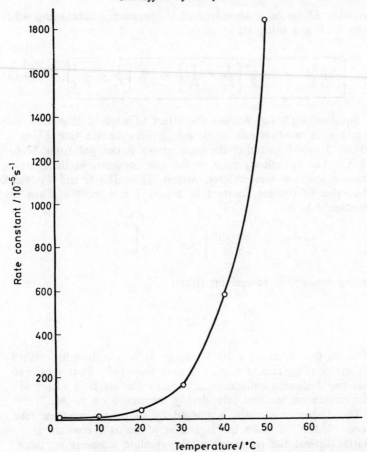

Figure 4.3 Decomposition of acetonedicarboxylic acid in aqueous solution (rate constant versus temperature)

Figure 4.3 shows the rate constant plotted directly against temperature from the same data. Examples of the calculations based on the Arrhenius equation are shown in Sections 13.5 and 13.6.

If the values of the rate constant at two temperatures T_1 and T_2 are k_1 and k_2, the energy of activation may be obtained from equation 4.9 as follows

$$\ln k_1 = \frac{-E^{\ddagger}}{RT_1} + \text{constant} \tag{4.11}$$

$$\ln k_2 = \frac{-E^{\ddagger}}{RT_2} + \text{constant} \tag{4.12}$$

assuming $E^{\ddagger}$ to be independent of temperature. Subtracting 4.12 from 4.11 and using M14

$$\ln \frac{k_1}{k_2} = -\frac{E^{\ddagger}}{R}\left[\frac{1}{T_1} - \frac{1}{T_2}\right] = \frac{E^{\ddagger}}{R}\left[\frac{1}{T_2} - \frac{1}{T_1}\right] \qquad (4.13)$$

Equation 4.13 also enables the effect of a 10 °C change in temperature on reaction rate to be calculated. Assume that $E^{\ddagger} = 80\,000$ J mol^{-1} and that the temperature is changed from 37 to 27 °C. Let k_1 and k_2 refer to the rate constants at the corresponding absolute temperatures, namely $T_1 = 310$ K and $T_2 = 300$ K The value of the gas constant R is 8.31 J K^{-1} mol^{-1}. Then from equation 4.13

$$\ln \frac{k_1}{k_2} = \frac{80\,000}{8.31}\left[\frac{1}{300} - \frac{1}{310}\right] = 1.035$$

Taking antilogs of natural logarithms

$$\frac{k_1}{k_2} = 2.82$$

Thus, in this instance, a 10 °C change in temperature has caused a corresponding rate change of almost threefold. It is thus seen that the Arrhenius equation can account for the large effect of temperature on reaction rate already mentioned on p. 34.

The Arrhenius equation is obeyed by all simple one-stage reactions. Indeed, if when the logarithm of the rate constant is plotted against the reciprocal of the absolute temperature, there is any marked deviation from a straight line, this is usually taken as an indication that the reaction is complex. The converse is not true. Kinetic results obtained from many complex reactions also show a good fit to the Arrhenius plot.

4.5 Typical Results

Examination of the Arrhenius equation in the exponential form (equation 4.10) shows that the constant A has the same dimensions as the rate constant. A and $E^{\ddagger}$ are referred to collectively as the Arrhenius parameters of a chemical reaction. Some values of A and $E^{\ddagger}$ in various reactions are given in *Table 4.1*.

Table 4.1

Reaction	Medium	$E^{\ddagger}/$ kJ mol^{-1}	$\log\left\{\dfrac{[A]/}{\text{l mol}^{-1}}\,\text{s}^{-1}\right\}$
$CH_3CO_2C_2H_5 + NaOH \rightarrow$	Water	47	7.2
$n\text{-}C_5H_{11}Cl + KI \rightarrow$	Acetone	77	8.0
$C_2H_5ONa + CH_3I \rightarrow$	Ethanol	82	11.4
$C_2H_5Br + NaOH \rightarrow$	Ethanol	90	11.6
$NH_4CNO \rightarrow NH_2CONH_2$	Water	97	12.6
$2N_2O_5 \rightarrow 2N_2O_4 + O_2$	Gas phase	103	13.7
$CH_3I + HI \rightarrow CH_4 + I_2$	Gas phase	140	12.2
$H_2 + I_2 \rightarrow 2HI$	Gas phase	165	11.2
$2HI \rightarrow H_2 + I_2$	Gas phase	167	10.7
$CH_3N_2CH_3 \rightarrow C_2H_6 + N_2$	Gas phase	220	13.5
$\begin{array}{c}CH_2{-}CH_2 \\ \backslash\,CH_2 \end{array} \rightarrow CH_3CH{=}CH_2$	Gas phase	272	12.2

In many second-order reactions the frequency factor A is about 10^{11} l mol^{-1} s^{-1}, i.e. $\log_{10} A$ is roughly 11. The reason for this is considered in the section dealing with the theory of reaction rates. Values of the activation energy range from 40 to 200 kJ mol^{-1} in most cases. The energy of activation is considerably less than the energy required to break the bond involved in the reaction because, in the activated state, the molecule has been stretched but the bond has not yet been broken.

4.6 Endothermic Reactions

In general, the higher the temperature required to bring about reaction, the higher is the energy of activation. This can be seen by taking logarithms of the Arrhenius equation 4.10 and using M11

$$\ln k = \frac{-E^{\ddagger}}{RT} + \ln A \qquad (4.14)$$

It will be seen that if the rate constant k or $\ln k$ is to have a specified value then $E^{\ddagger}/RT$ must be constant. In other words if two reactions are to proceed at the same rate, $E^{\ddagger}/RT$ for each reaction must have the same value. Thus the reaction requiring a high temperature is the one with a high energy of activation. In practice A does vary to some extent from reaction to reaction, so this deduction is not strictly quantitative. Nevertheless, high energies of activation are associated with reactions requiring high temperatures.

It is also worth noting that if a reaction is highly endothermic (i.e. occurs with a large absorption of heat from the surroundings), it will have a large energy of activation and will therefore require a high temperature. This follows from equation 4.3 where it is shown that

$$\Delta H = E_1^{\ddagger} - E_{-1}^{\ddagger}$$

$E_1^{\ddagger}$ and $E_{-1}^{\ddagger}$ are the energies of activation of the forward and back reaction, respectively, and ΔH is the heat of reaction. $E_1^{\ddagger}$ and $E_{-1}^{\ddagger}$ are necessarily positive, so that if ΔH is positive, as it is in an endothermic reaction

$$E_1^{\ddagger} = \Delta H + E_{-1}^{\ddagger}$$

i.e.

$$E_1^{\ddagger} > \Delta H$$

Hence it follows that a strongly endothermic reaction will have a large energy of activation and will therefore require a high temperature to bring about the reaction at a measurable rate.

If a mechanism of reaction can be found which will provide a path of lower energy of activation, there will be a corresponding increase in the rate of reaction. This is in fact how catalysts function. They provide an alternative reaction path of lower energy of activation, but are not themselves consumed in the process.

4.7 The Arrhenius Equation and the Boltzmann Factor

The Arrhenius equation was originally formulated by using the van't Hoff isochore (equation 4.4) and assuming that the equilibrium constant is the ratio of the rate constants of the forward and back reactions. There is, however, an alternative way in which equation 4.10 can be obtained.

Arrhenius postulated that only activated molecules could bring about reaction, and hence at constant temperature it would be expected that the rate of reaction should be proportional to the fraction of molecules which were activated. Suppose that the concentration of each reactant is kept constant at one mole per litre. Then the rate of reaction is equal to the rate constant (k) (see Section 1.4) and so k is proportional to the number of activated collisions or (using M1)

$$k = A \times \text{(fraction of molecules with activation energy)} \quad (4.15)$$

where A is a proportionality constant having the same dimensions as k.

From the Maxwell–Boltzmann theory of the distribution of energy among molecules it is known that the fraction of molecules possessing an energy E in excess of the average energy is equal to $\exp(-E/RT)$. This exponential term is known as the Boltzmann Factor. For very high energies of activation the value tends to zero. For small or zero energies of activation the value tends to unity. The Boltzmann factor therefore always has values ranging between zero and unity. It therefore follows from equation 4.15 that

$$k = A \exp\left(\frac{-E}{RT}\right) \qquad (4.16)$$

which is the Arrhenius equation in the exponential form stated in equation 4.10.

For most reactions $\exp(-E/RT)$ is a very small fraction. Taking as a typical example a reaction at $25\,°C$ in which the energy of activation is 80 kJ mol^{-1}

$$\exp\left(\frac{-E}{RT}\right) = \exp\frac{(-80\,000)}{8.31 \times 298} = \exp(-32.30) \approx 10^{-14}$$

i.e. the fraction of collisions which are activated is 10^{-14}. It is seen, therefore, that one collision in every 10^{14} is activated.

The population of the world is approximately four thousand million or 4×10^9. Thus the fraction of activated molecules is about one hundred thousand times smaller than the fraction of the world's population which is reading this sentence at this very moment! The slow rate of chemical reactions with this activation energy can now be understood, for before reacting the average molecule will require 10^{14} collisions to become activated. Even though the collision frequency is as high as 10^{11} per second per molecule, the time taken for a molecule to react is

$$\frac{10^{14}}{10^{11}} = 1000 \text{ s}$$

The above discussion shows the relation between the kinetic theory of gases, as developed by Maxwell and Boltzmann, and the kinetics of chemical reactions as developed by Arrhenius. This relationship is brought out more clearly by examining *Figure 4.4*, which shows in a diagram the Maxwell–Boltzmann law applied to the distribution of molecular velocities at various temperatures.

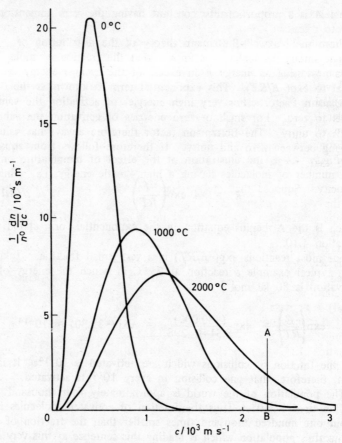

Figure 4.4 Maxwell–Boltzmann distribution law

The Maxwell–Boltzmann distribution law is usually expressed, for a gas, as

$$\frac{1}{n_0}\frac{dn}{dc} = 4\pi c^2 \left(\frac{m}{2\pi kT}\right)^{3/2} \exp\left(\frac{-mc^2}{2kT}\right) \qquad (4.17)$$

where dn is the number of molecules with velocities between c and $c + dc$, m is the mass of the molecule, k is the Boltzmann constant, T is the absolute temperature and n_0 is the total number of molecules present.

Equation 4.17 is a somewhat formidable expression, but its usefulness may be more easily seen if it is realized that the right-hand side is only a function of m, T and c, the other terms being constants. Thus for a gas of known molecular weight it is possible to calculate dn/n_0 given the values of T and c. This is how a diagram of the type shown in *Figure 4.4* is prepared.

The reason for choosing $dn/n_0 dc$ as ordinate is that it gives direct meaning to the area under the curve. The area between two velocities c_1 and c_2 will be the fraction of the molecular population lying between these two velocities. (The total area under the curve must necessarily be unity, since all the molecules have velocities between zero and infinity.)

From the viewpoint of chemical kinetics, the important aspect of *Figure 4.4* is the illustration of the effect of temperature on the number of molecules having a high kinetic energy, i.e. a high velocity. Suppose that the vertical line AB in *Figure 4.4* is drawn at the velocity corresponding to the activation energy for a reaction of the molecules. Then the area under the Maxwell–Boltzmann curve to the right of AB represents the fraction of the molecules that are activated, i.e. that possess more than the activation energy. It is seen by inspection that at $0\,^{\circ}C$ only a minute fraction is activated. At $1000\,^{\circ}C$ the fraction activated is just noticeable, whereas at $2000\,^{\circ}C$ a much larger fraction (approximately one tenth of the total) is activated.

It is not commonly realized that the somewhat complicated expression for the Maxwell–Boltzmann distribution law shown in equation 4.17 is simplified if the distribution is described in terms of the ratio

$$Q = \frac{\text{actual velocity of the molecules}}{\text{most probable velocity of the molecules}}$$

Using the result that the most probable velocity is $(2kT/m)^{\frac{1}{2}}$, which is the velocity corresponding to the maximum of each curve shown in *Figure 4.4*, then 4.17 simplifies to

$$dF/dQ = 4Q^2 \exp(-Q^2)\,/\pi^{\frac{1}{2}} \qquad (4.18)$$

where dF is the fraction of the molecules with velocity between Q and $Q + dQ$. The relative simplicity of 4.18 compared to 4.17 may make it easier for the reader to grasp the quantitative aspects of the Maxwell–Boltzmann distribution law.

The idea that the rate of a process is governed by an energy barrier or activation energy is one of the fundamental concepts of physical chemistry. It is also applied to problems outside the scope of chemical kinetics, e.g. viscosity of liquids, conductivity of semiconductors and decay of radioactive material.

4.8 Physical Properties and Chemical Reactivity

One of the difficulties of theoretical chemistry is that there is no correlation between physical properties and rates of reaction. The concept of activation energy helps to explain this result.

The average, and hence unactivated, molecules play a predominant role in determining the values of physical and thermodynamic properties, but *average molecules take no part in chemical reactions.* The activated molecules which participate in reaction are present to such a small extent that they have no influence on the average properties. There is, therefore, no correlation between, say, free energy changes and rates of reaction. Free energy values can only be related to *equilibrium conditions* (e.g. by the van't Hoff isotherm $\Delta G^{\ominus} = -RT \ln K$).

The chemical reactivity is related to the activation energy of the process which is determined by factors other than merely the initial and final states of the system, which is all that is needed, say, to calculate the free energy or enthalpy change of the chemical reaction.

5

THEORY OF REACTION RATES

5.1 General Points

The rates of chemical reactions range through the extremes of being imperceptibly slow to being virtually instantaneous. Any theory of reaction rates must explain why this is so.

The approach of Arrhenius was essentially empirical, but provided a basis from which more theoretical models were developed. The parameters A and E^+ of the Arrhenius equation for any particular elementary reaction must relate very closely to the specific requirements of the reacting species. Thus the theoretical models make their approach from a consideration of these requirements. Necessarily, therefore, the models are fundamentally related but emphasize different features.

No universal theory of reaction rates has as yet evolved from which it is possible to predict reaction rates for reactions in general. However, continued attempts to reconcile operational values with theoretical models form part of the effort towards deepening our understanding of chemical reactions. For certain elementary systems involving diatomic or triatomic molecules in the gas phase it has been possible to calculate the magnitude of the rate constant from first principles. For the more intractable reaction systems, such as reactions in solution, the chief value of reaction rate theory is to provide a standard of molecular behaviour as a yardstick against which actual experimental behaviour is compared.

5.2 Outline of Theoretical Approaches

There are two main theoretical approaches. The first is known as the *Collision Theory of Reaction Rates* and is based on the kinetic theory of gases. It applies chiefly to bimolecular reactions starting from the premise that if two molecules are to combine chemically an essential first step is that they should meet (i.e. collide). Then, using Arrhenius's concept of an activation energy, it is postulated that not all collisions lead to reaction. Only those in which the molecules acquire more than the activation energy will be able to do so. The collision theory may therefore be expressed in a sentence by saying that the rate of reaction is equal to the number of activated collisions per unit time.

The second method of approach, known as the *Transition State Theory*, is based upon a more detailed consideration of the concept of a 'collision leading to reaction'. Instead of regarding the reaction of molecules as something which either happens instantaneously or does not happen, it envisages that the bond-breaking and bond-making involved in a chemical change must occur continuously and simultaneously. This idea can be expressed in a potential energy diagram. Consider the potential energy changes as reactant molecules pass through a transition state to become product molecules, as shown in *Figure 5.1*.

The reactant molecules are initially at a potential energy minimum. On receiving the required collision energy, the reactant molecules co-operate in a concerted process in which the energy imparted by the collision is distributed in bond-stretching and incipient bond-making. Thus at the expense of the energy of collision, the potential energy increases steadily until the transition state is reached.

The molecular entity existing at the potential energy maximum must be held together by abnormally long bonds in a high state

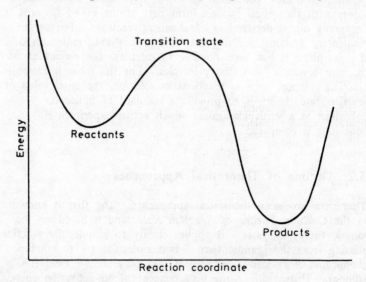

Figure 5.1   Potential energy changes in a chemical reaction

of strain. After a transitory life, the entity bursts apart, either to complete fully the new bonds of the product molecules, or to restore the initial bonds and revert to reactant molecules.

The *reaction co-ordinate* is a convenient parameter with which to assess the extent of the reaction. It represents a point on the potential energy diagram which is imagined to move continuously from the initial to the final equilibrium state. *Figure 5.1* should be compared with the simpler diagram shown in *Figure 4.1* on p. 35.

The concept of a transition state can be visualized more clearly by considering a definite example such as the elementary reaction between molecular hydrogen and atomic chlorine. This is a transfer process in which there is one bond forming in exchange for one bond breaking up:

$$H_2 + Cl \rightarrow H + HCl$$

Thus if a partially formed or partially broken bond is denoted by a dotted line, the reaction may be written as

$$H-H + Cl \rightarrow H \cdots H \cdots Cl \rightarrow H + HCl \qquad (5.1)$$

The intrinsically unstable entity with the elongated bonds is known as the *Transition Complex* or the *Activated Complex*. The rise in potential energy that is required for reactant molecules to form the transition complex corresponds to the activation energy of the forward-proceeding process.

The transition state approach, sometimes referred to as the *Theory of Absolute Reaction Rates*, assigns thermodynamic equilibrium properties to the transition complex, although the species has only a brief existence (about 10^{-13} s) and cannot be isolated. It is assumed that the process of moving from the separate reactant molecules to the associated condition of the transition state is reversible. Therefore there is effectively an equilibrium between the reactants and the transition state and so normal thermodynamic methods can be applied. In principle this allows the rate equation to be predicted from the properties of the reactants. These two theoretical approaches will now be described in more detail.

5.3 Essentials of Collision Theory

The basic idea of this theory is that the rate of reaction equals the number of activated collisions per unit time.

Consider now the reaction between the gases A and B whose molecules are regarded as hard spheres in chaotic motion. From the Maxwell–Boltzmann theory of the distribution of energy among molecules (Section 4.7), it is known that the fraction of molecules having energy greater than or equal to the activation energy is $\exp(-E^{\ddagger}/RT)$. Assuming the reaction is bimolecular of the form

$$A + B \rightarrow AB$$

then

$$\frac{dn}{dt} = Z_{AB} \exp\left(\frac{-E^{\ddagger}}{RT}\right) \tag{5.2}$$

where n = number of molecules of product formed per unit volume

Z_{AB} = number of collisions between A and B per unit volume per unit time

Z_{AB} and dn/dt are not constants since the number of reactant molecules steadily decreases as the reaction proceeds.

It follows by definition of the second-order rate constant k that

$$\frac{dn}{dt} = k n_A n_B \tag{5.3}$$

where n_A and n_B are the number of molecules per unit volume of A and B, respectively, and k is expressed in the appropriate units. Combining 5.2 and 5.3

$$k = \frac{Z_{AB}}{n_A n_B} \exp\left(\frac{-E^{\ddagger}}{RT}\right)$$

or

$$\boxed{k = Z \exp\left(\frac{-E^{\ddagger}}{RT}\right)} \tag{5.4}$$

where

$$Z = \frac{Z_{AB}}{n_A n_B} \tag{5.5}$$

and is called the *collision constant* or *collision number*. Equation 5.4 is a mathematical statement of the Collision Theory of Reaction Rates. It should be noted that the exponential form of the

Arrhenius equation 4.10 is identical with 5.4 if the constant A is replaced by Z.

From the kinetic theory of gases, the collision constant is proportional to the square root of the absolute temperature, and so there are two temperature-dependent terms in the expression for k in 5.4. However, the increase in Z with temperature is so small compared with the change in the exponential term that it is usually ignored.

For example, at 27 °C (300 K) a 10 °C rise in temperature will increase Z by a factor of $(310/300)^{1/2} = 1.016$. This may be compared with a two- to three-fold increase in the exponential term for the same temperature rise. The slight dependence of Z on temperature is not sufficient to upset the linearity of the Arrhenius plot of $\ln k$ against $1/T$, but it must be allowed for in accurate measurements of the energy of activation.

Verification of the collision theory consists of calculating the collision constant (Z), and measuring the energy of activation ($E^{\ddagger}$) for a particular reaction. These values are substituted in the expression $Z \exp(-E^{\ddagger}/RT)$ and the result compared with the experimental value of the rate constant (k).

5.4 Calculation of the Collision Constant

From the kinetic theory of gases it is known that, in a mixture of gases A and B, the number of collisions (Z_{AB}) involving one molecule of A and one molecule of B taking place in one cubic metre in one second is

$$Z_{AB} = n_A n_B \sigma_{AB}^2 [8\pi RT(M_A + M_B)/M_A M_B]^{1/2} \qquad (5.6)$$

where M_A and M_B are the molar masses of A and B respectively and σ_{AB} is the mean collision diameter of A and B.

The collision diameter of a single molecule is the diameter of the effective target area that it presents to other colliding molecules, assuming that the molecules are spherical. The mean collision diameter of A and B is the average of their individual collision diameters, i.e.

$$\sigma_{AB} = \tfrac{1}{2}(\sigma_A + \sigma_B)$$

Substitution of the collision constant $Z = Z_{AB}/n_A n_B$ (from 5.5) into 5.6 gives

$$Z = \sigma_{AB}^2 [8\pi RT(M_A + M_B)/M_A M_B]^{1/2} \qquad (5.7)$$

The physical significance of Z is that it represents the number of collisions per unit volume per unit time when the reactants are at unit concentration. This may be seen by substituting $n_A = n_B = 1$ into 5.5, giving $Z = Z_{AB}$.

If the reaction involves only the molecules of a single gas, e.g.

$$2HI \rightarrow H_2 + I_2$$

then $M_A = M_B$ and 5.7 reduces to

$$Z = 4\sigma^2 (\pi RT/M)^{\frac{1}{2}} \tag{5.8}$$

The collision diameter (σ) of a gas can be estimated from data on the viscosity (η) and density (ρ) of the gas using the formula derived from the kinetic theory of gases

$$\sigma^2 = \frac{2\rho}{3\pi\eta n} \left[\frac{RT}{\pi M} \right]^{\frac{1}{2}} \tag{5.9}$$

The collision diameter of most gas molecules is in the range of 0.1–1.0 nm. Some values of collision diameters from viscosity measurements are shown in *Table 5.1*.

Table 5.1

Gas	Collision diameter/nm	Gas	Collision diameter/nm
Hydrogen	0.25	Chlorine	0.45
Helium	0.22	Hydrogen iodide	0.35
Hydrogen chloride	0.29	Carbon dioxide	0.42
Nitrogen	0.35	Mercury	0.63
Oxygen	0.34	Benzene	0.75

It should be noted that equations 5.7 and 5.8 are derived by assuming that the molecules are spherical. This is certainly not the case for HCl, HI, CO_2 and others, and so the collision diameters in these cases must be regarded as mean values.

5.5 Calculation of a Rate Constant using Collision Theory

It must be emphasized that the collision theory does not allow a rate constant to be predicted from a knowledge of the properties of the reactants and products. It provides a model against which measured properties are tested. Thus the collision diameter is estimated from properties such as the gas viscosity, and then the collision constant Z is calculated. The energy of activation $E^{\ddagger}$ is determined from the variation of reaction rate with temperature. The values of Z and $E^{\ddagger}$ are substituted into equation 5.4 and a collision theory value of the rate constant k is obtained. This result is compared with the value of the rate constant determined by direct measurement of the rate of reaction. Agreement within an order of magnitude is sometimes observed, although compensating factors may give a closer fit than is justified by the simple collision theory. Abnormally large discrepancies serve to indicate interesting anomalies that deserve further examination using other treatments.

A reaction for which there is good agreement with the collision theory is

$$HI \rightarrow \tfrac{1}{2}H_2 + \tfrac{1}{2}I_2$$

Data for this reaction are used to illustrate the method of calculation. From the variation of reaction rate with temperature an energy of activation of 184 kJ mol^{-1} is obtained. The collision diameter of the hydrogen iodide molecule, estimated from the viscosity of the gas, is 0.35 nm. Thus the collision constant calculated from equation 5.8 is

$$Z = 4\sigma^2 [\pi RT/M]^{1/2} = 1.651 \times 10^{-16} \text{ m}^3 \text{ s}^{-1}$$

Particular care must be taken about using consistent units. In this calculation, the following values are appropriate:

$E^{\ddagger} = 184$ kJ mol^{-1}

$\sigma = 3.5 \times 10^{-10}$ m

$T = 556$ K

$R = 8.314$ J K^{-1} mol^{-1}

$M = 127.9 \times 10^{-3}$ kg mol^{-1}

From equation 5.8 the dimensions of Z are

$$\text{m}^2 \left(\frac{\text{J K}^{-1} \text{ mol}^{-1} \text{ K}}{\text{kg mol}^{-1}} \right)^{1/2}$$

Substituting for J the dimensions kg m^2 s^{-2}, the dimensions of the collision constant reduce to m^3 s^{-1}.

The second-order rate constant is usually expressed in l mol^{-1} s^{-1}. The conversion factor between the two sets of units will therefore be 1000 N where N is the Avogadro Constant $(6.02 \times 10^{23}$ mol$^{-1})$. Hence

$$Z = 1.651 \times 10^{-16} \times 6.02 \times 10^{23} \times 1000$$

$$= 9.93 \times 10^{10} \text{ l mol}^{-1} \text{ s}^{-1}$$

This value of the collision constant and the activation energy are then substituted into equation 5.4, namely

$$k = Z \exp(-E^{\ddagger}/RT)$$

to give

$$k = 9.93 \times 10^{10} \exp\left(\frac{-184\,000}{8.314 \times 556}\right)$$

$$= 5.2 \times 10^{-7} \text{ l mol}^{-1} \text{ s}^{-1}$$

This collision theory value is in good agreement with the experimentally determined value of 3.5×10^{-7} l mol^{-1} s^{-1}.

It is interesting to note that for a long time the second-order thermal decomposition of hydrogen iodide was regarded as a single stage bimolecular reaction and its reverse reaction an exemplary bimolecular collision process. However, in 1967 J.H. Sullivan showed that the mechanism is undoubtedly stepwise. He concluded that the reverse reaction between hydrogen and iodine could proceed through the rapid equilibrium

$$I_2 \rightleftharpoons 2I\cdot$$

followed by a termolecular rate-controlling stage

$$H_2 + 2I\cdot \rightarrow 2HI$$

Symmetry rules for four-centre reactions of diatomic molecules also forbid the single-stage collision mechanism

$$\begin{array}{ccc} H & H & H-H \\ | & + \; | & \rightleftharpoons \quad + \\ I & I & I-I \end{array}$$

.6 Extension of Collision Theory to Reactions in Solution

he collision constant in gas reactions has been calculated in quation 5.7 from the kinetic theory of gases by assuming that e frequency of collision is that of incompressible spherical mole-les which exert no forces on one another and are not influenced y the presence of other molecules. It therefore seems possible at in ideal solutions the frequency of collisions should approxi-ate to that calculated by the gas method.

This approximation is justified by the fact that some reactions ccur at the same rate in solution as in the gas phase, examples eing the decomposition of dinitrogen pentoxide and the reaction of hlorine with ozone. Supporting evidence is also found from the ct that reactions in solution often have a frequency factor close) that calculated by the collision theory. Examples are shown a *Table 5.2*.

able 5.2

Reactants	*Solvent*	$\dfrac{E}{\text{kJ mol}^{-1}}$	$\log\left\{\dfrac{A(\text{observed})}{\text{l mol}^{-1}\,\text{s}^{-1}}\right\}$	$\log\left\{\dfrac{Z(\text{calculated})}{\text{l mol}^{-1}\,\text{s}^{-1}}\right\}$
$_2H_5ONa + CH_3I$	Ethanol	82	11.3	11.2
$_2H_5Br + OH^-$	Ethanol	90	11.6	11.5
$CH_3)_2SO_4 + KCNS$	Methanol	75	10.7	11.2

'here are many reactions in solution, particularly between an ion nd a polar molecule, where log A is between 11 and 12 (the value o be expected from collision theory).

.7 Limitations of Simple Collision Theory

Although the collision theory can account for the rate of some eactions, many are known where the rates differ by several powers f ten from the values calculated from equation 5.4. For this eason the equation is often written

$$k = PZ \exp\left(\frac{-E^{\ddagger}}{RT}\right) \qquad (5.10)$$

vhere P (called the *probability* or *steric* factor) is a correcting erm, which can be regarded as the ratio of the observed rate con-tant to that calculated from the collision theory.

The reaction between triethylamine and ethyl iodide in the vapour phase occurs more slowly than expected ($P = 10^{-8}$), whereas the acid hydrolysis of some sugars occurs faster than expected with values of P up to 10^9.

Clearly the simple collision theory is not the right model for interpreting these reactions. Some of the inadequacies of the simple collision theory can be attributed to the following.

(a) Because bimolecular gas-phase reactions involving large molecules often have P values much smaller than unity, it is contended that polyatomic reactants may need a particular orientation before they can react. Views differ on the order of magnitude of the contribution of this requirement. However, it is considered unlikely that this would result in P values as low as 10^{-5}, as for example is found for the reaction

$$CH_2 = CHCH = CH_2$$

$$+$$

$$CH_2 = CHCHO$$

$$\rightarrow$$

$$
\begin{array}{c}
H_2 \\
C \\
HC \diagup \quad \diagdown CHCHO \\
\| \qquad\qquad | \\
HC \diagdown \quad \diagup CH_2 \\
C \\
H_2
\end{array}
$$

where the Arrhenius parameter A is about 10^6 1 mol^{-1} s^{-1}.

Similarly, the presence of large groups or particular structures may shield a position normally vulnerable to indiscriminate energetic collision. In such a case the energy of activation is probably raised rather than the number of effective collisions being substantially reduced. Thus very low P values cannot be explained satisfactorily by simple collision theory.

The over-simplified hard-sphere model disregards any of the internal contributions made by the vibrational and rotational states of the molecules. By allowing for these factors the order of magnitude of the factor P may be explained.

(b) Reactions may occur, not in a single step, but by a chain mechanism in which a molecule of product is capable of further reaction (see Chapter 11). This will cause high P factors.

(c) If the reaction occurs on a catalytic surface the collision theory cannot be applied since the rate of reaction does not depend on the number of collisions in the gas (see Chapter 8).

(d) Another aspect neglected in the simple theory of colliding hard spheres is the duration of contact. The concept of an instantaneous bimolecular collision takes no account of short-range forces of interaction between reactant molecules. However, at ordinary temperatures it is estimated that the probable time during which these forces are appreciable is about 10^{-13} s, which is of the same order of magnitude as the period of a molecular vibration.

In processes involving two simple molecules or atoms, where a single product molecule forms, an excess of energy imparted by the collision must be removed within the period of a single vibration; otherwise dissociation into initial reactants is inevitable. During the 10^{-13} s interval, if a third body is encountered, the excess energy can be transferred so that there is a stable resultant product. The simple hard-sphere model implies a zero time of contact and excludes the possibility of the second stabilizing collision.

Summing up, the collision theory gives a satisfactory account of the rates of some of the simpler reactions, but needs major modifications if it is to be more widely applicable. In particular, the concept of collision leading to reaction, developed with the transition state approach, allows for potential energy changes during the time of effective collision. This cannot be done with the simple collision theory model.

5.8 Transition State Theory

The main assumption of this theory is that all chemical reactions proceed via a transition state which is in thermodynamic equilibrium with the reactants even though the overall chemical reaction is irreversible. This may be symbolized as

$$A + B \rightleftharpoons C^{\ddagger} \rightarrow \text{Products}$$

where the superscript ‡ refers to the transition state.

The rate of reaction is assumed to be proportional to the concentration of the activated molecules $(c^{\ddagger})$ and this concentration is governed by the laws of chemical equilibria. Thus if $K^{\ddagger}$ refers to the equilibrium constant for formation of the transition state, and a, b and $c^{\ddagger}$ represent the concentrations of A, B and $C^{\ddagger}$ at time t, then

$$K^{\ddagger} = c^{\ddagger}/ab \text{ or } c^{\ddagger} = K^{\ddagger}ab \qquad (5.11)$$

The rate of reaction is proportional to the concentration of the transition complex $(c^{\ddagger})$, or

$$\frac{dx}{dt} = mc^{\ddagger} \qquad (5.12)$$

where m is a proportionality constant having the dimensions of frequency, for example s^{-1}. However, by the definition of a second-order rate constant (k), (see equation 1.8)

$$\frac{dx}{dt} = kab \qquad (5.13)$$

Equating 5.12 with 5.13

$$k = \frac{mc^{\ddagger}}{ab} = mK^{\ddagger} \text{ (from 5.11)} \qquad (5.14)$$

Remembering that the transition complex is a highly strained state between bond-forming and bond-breaking, the constant m can be interpreted as the frequency with which the bond about to break vibrates to dissociation. Thus the two product-forming ends of the bond can be visualized as moving away in opposite directions, each having obtained, in classical terms, the average kinetic energy of translation, namely $RT/2N$. Hence the frequency of vibration v of the disintegrating bond can be calculated using the Planck relationship

$$\Delta E = hv = RT/N$$

where ΔE is the bond dissociation energy.

The theory considers m to be equal to the frequency of the abnormally strained vibration in the transition complex. So equation 5.14 becomes

$$k = \frac{RTK^{\ddagger}}{Nh} \qquad (5.15)$$

The factor RT/Nh is the same for all molecules irrespective of their chemical properties. Its value is determined by the temperature. For example at 27 °C

$$T = 300 \text{ K}$$

The values of the other constants are

$$R = 8.31 \text{ J K}^{-1} \text{ mol}^{-1}$$

$$N = 6 \times 10^{23} \text{ mol}^{-1}$$

$$h = 6.6 \times 10^{-34} \text{ J s}$$

so that

$$\frac{RT}{Nh} = m = \frac{8.31 \times 300}{6 \times 10^{23} \times 6.6 \times 10^{-34}}$$

$$= 6.3 \times 10^{12} \text{ s}^{-1}$$

which is the vibrational frequency of a fairly weak bond.

At this point it is also appropriate to consider the dimensions of the classical equilibrium constant $K^{\ddagger}$, as defined in equation 5.11, which is derived by assuming a bimolecular reaction. $K^{\ddagger}$ has the dimensions of 1 mol^{-1}. The right-hand side of 5.15 has dimensions of $1 \text{ mol}^{-1} \text{ s}^{-1}$, which is consistent with the dimensions of a second-order rate constant. However, it should be noted that the equilibrium constant $K^{\ddagger}$ when used in thermodynamic relationships must be dimensionless. This discussion will be raised again in Section 14.3.

5.9 Thermodynamics and the Rate Equation

The transition state theory enables the methods of thermodynamics to be applied to the kinetics of chemical reactions by making the assumption that *there is a thermodynamic equilibrium between the reactants and the transition complex*. This section shows how the conclusions of the theory can be found starting from standard thermodynamic results. The derivation of these results from the laws of thermodynamics and some explanation of the concept of activity are given in Section 14.3.

The thermodynamic result, on which the transition state theory is based, is

$$\Delta G^{\ddagger} = \Delta H^{\ddagger} - T\Delta S^{\ddagger} = -RT \ln K^{\ddagger} \tag{5.16}$$

where $\Delta G^{\ddagger}$ is the standard free energy change of formation of the transition state

$\Delta H^{\ddagger}$ is the standard enthalpy change of formation of the transition state (enthalpy of activation)

$\Delta S^{\ddagger}$ is the standard entropy change of formation of the transition state (entropy of activation)

From 5.16

$$\ln K^{\ddagger} = \frac{T\Delta S^{\ddagger} - \Delta H^{\ddagger}}{RT} \tag{5.17}$$

Using M9 and M10

$$K^{\ddagger} = \exp\left(\frac{\Delta S^{\ddagger}}{R}\right) \exp\left(\frac{-\Delta H^{\ddagger}}{RT}\right) \tag{5.18}$$

If this result is substituted into 5.15

$$k = \frac{RT}{Nh} \exp\left(\frac{\Delta S^{\ddagger}}{R}\right) \exp\left(\frac{-\Delta H^{\ddagger}}{RT}\right) \tag{5.19}$$

Equation 5.19 is the expression for the rate constant of a reaction derived from the combination of thermodynamic results and transition state theory. It is useful to compare this with the corresponding equation 5.10 from collision theory, namely

$$k = PZ \exp\left(\frac{-E^{\ddagger}}{RT}\right)$$

The terms $\exp(-\Delta H^{\ddagger}/RT)$ and $\exp(-E^{\ddagger}/RT)$ in 5.19 and 5.10 are almost the same, since $\Delta H^{\ddagger}$ represents the increase in standard enthalpy on forming the transition state from the reactants, and $E^{\ddagger}$ is the experimental activation energy (see Section 4.2). The difference between these two energies is usually quite small.

The significant difference between 5.10 and 5.19 is that the term PZ of the collision theory is now replaced by $(RT/Nh) \exp(\Delta S^{\ddagger}/R)$. RT/Nh has a value of about 10^{13} s^{-1} at 25 °C, whereas Z (which is proportional to the square root of T) is approximately 10^{11} 1 mol^{-1} s^{-1} at 25 °C. The difference in dimensions and the relation to temperature of the pre-exponential terms in the collision theory and transition state theory equations for the rate constant should be noted.

The term $\exp(\Delta S^{\ddagger}/R)$ enables deviations from the collision theory to be interpreted in terms of entropy (or order–disorder) changes in forming the transition state. Thus if a reaction requires a specific orientation of the reactants, the transition state is more ordered than the initial state, then $\Delta S^{\ddagger}$ is negative and $\exp(\Delta S^{\ddagger}/R)$ is less than unity. This means that the rate of reaction predicted by the transition state theory will be less than that predicted on simple collision theory, which is in fact the case. It corresponds to a value of P smaller than unity.

The bimolecular gas reaction

$$2CH_2=CHCH=CH_2 \rightarrow$$

results in a loss of vibrational freedom when the two polyatomic molecules form the single activated complex. Thus the value of $\Delta S^{\ddagger}$ will be negative. The experimental value is -54 J K^{-1} mol^{-1}

Conversely, if the transition state is less ordered than the initial state, $\Delta S^{\ddagger}$ will be positive, corresponding to a high value of *P*. This occurs in unimolecular gas reactions in which the reactant molecule forms a transition complex with a gain in vibrational freedom. For example, in the reaction

$$\begin{array}{c} H_2 \\ C \\ H_2C \underline{\hspace{1cm}} CH_2 \end{array} \rightarrow CH_3CH=CH_2$$

the value of $\Delta S^{\ddagger}$ is 46 J K^{-1} mol^{-1} at 25 °C.

Quite small changes in entropy produce a large kinetic effect. For example, if $\Delta S^{\ddagger} = 83$ J K^{-1} mol^{-1}, then

$$\frac{\Delta S^{\ddagger}}{R} \approx 10$$

Therefore

$$\exp\left(\frac{\Delta S^{\ddagger}}{R}\right) = \exp(10) \approx 2 \times 10^4$$

Thus the value of *P* found by using the collision theory would be about 10^4. An entropy change of 160 J K^{-1} mol^{-1} accounts for *P* factors of 10^8. It may help to appreciate whether such entropy changes are large or small by molecular standards to realize that for the common gases at STP the molar entropy is in the region of 200 J K^{-1} mol^{-1}. If a reaction has a similar mechanism in the gas phase and in solution, then provided that solvation is not involved, $\Delta H^{\ddagger}$ and $\Delta S^{\ddagger}$ should not vary significantly with the medium. In this way it can be explained why some reactions occur at the same rate in solution and in the gas phase.

It is now possible to give some account of the widespread variation in rate from reaction to reaction in terms of two factors, namely the *energy* and *entropy* of activation. In general the energy of activation will depend on the strength of the bonds that are being broken and formed in the transition state, but it will also be considerably influenced by factors such as solvation. The value may vary from 40 to 320 kJ mol^{-1}. At room temperature this variation in energy of activation would give rate constants which differed by a factor of 10^{50}. *The energy of activation is the major factor in determining the rate of reaction.*

The entropy term ($\Delta S^{\ddagger}$) is not so easy to visualize as is the energy of activation. However, it plays a profound part in reactions in which orientation of molecules is involved (as is also seen in the next section). Its magnitude is sufficient in extreme cases to give rates of reaction which differ by a factor of 10^{10}.

One purpose of the theory of reaction rates is to interpret the rates of chemical reactions from knowledge of molecular structure. Simple collision theory is of limited use for this purpose as quantitative significance cannot be attached to the probability factor P. However, the transition state theory shows (equation 5.19) that the problem of predicting rates of reaction is one of predicting energies and entropies of activation. Attempts have been made to do this by combining the results of quantum and statistical mechanics. This approach is not discussed in this text except to say that in some simple cases, such as the conversion of para-hydrogen to ortho-hydrogen by atomic hydrogen, reasonable results have been obtained.

5.10 Effect of Pressure on Reaction Rates in Solution

The effect of pressure on reaction rates has been extensively investigated in recent years. To obtain measurable effects on reactions in solution, pressures of thousands of atmospheres must be applied. Instrumental methods of analysing the reaction mixture whilst it is under pressure are also needed. This approach has greatly strengthened the power of the kinetic method of investigating reaction mechanisms.

The transition state theory enables a simple explanation to be given of the effect of pressure on the rate constants of reactions in incompressible solutions, i.e. in those reactions where an increase in pressure does not decrease the total volume of the reaction mixture. The basic idea is derived from equation 5.14 in which it is shown that the *rate constant of a reaction is proportional to the equilibrium constant for the formation of the transition state*. By conventional thermodynamic methods it is possible to calculate the effect of pressure on equilibrium constants, which, therefore, gives the effect of pressure on the rate constant.

To investigate the effect of pressure at constant temperature, the van't Hoff isotherm

$$-\Delta G^{\ddagger} = RT \ln K^{\ddagger}$$

is partially differentiated with respect to pressure giving

$$-\left[\frac{\partial(\Delta G^\ddagger)}{\partial p}\right]_T = -\Delta V^\ddagger = RT\left[\frac{\partial(\ln K^\ddagger)}{\partial p}\right]_T \quad (5.20)$$

In obtaining the middle term of 5.20, use has been made of the thermodynamic result

$$\left(\frac{\partial G}{\partial p}\right)_T = V \quad (5.21)$$

The quantity $\Delta V^\ddagger$ in 5.20 is called the *volume of activation*. It represents the change in volume of the whole system on forming the transition state from the reactants. The volume of activation includes volume changes due to changes in the type of solvation. Consequently, the volume of activation can be either positive or negative.

By assuming that the system is incompressible, i.e. that the volume of activation does not depend on pressure, then integrating 5.20 at constant temperature gives

$$\ln K^\ddagger = -p\Delta V^\ddagger/RT + \text{constant} \quad (5.22)$$

From 5.14 the rate constant (k) is proportional to the equilibrium constant $(K^\ddagger)$. Taking logarithms of 5.14 and using M13 gives

$$\ln k = \ln K^\ddagger + \ln m \quad (5.23)$$

where m is a constant. Combining 5.22 and 5.23

$$\ln k = -p\Delta V^\ddagger/RT + m' \quad (5.24)$$

m' being a new unknown constant. If k_0 is the rate constant when the applied pressure is zero, then $m' = \ln k_0$. Hence

$$\ln k = \ln k_0 - p\Delta V^\ddagger/RT \quad (5.25)$$

Thus, according to the transition state theory, in any incompressible system *the logarithm of the rate constant will vary linearly with the applied pressure*. This result has been verified in many experiments in which hydraulic pressures of a few thousand atmospheres have been applied to the reaction system. The rates of reaction are followed by an instrumental method, such as measurement of electrical conductance.

Using M14, equation 5.25 can be rearranged to give

$$\boxed{\ln(k/k_0) = -p\Delta V^\ddagger/RT} \quad (5.26)$$

This requires that the plot of $\ln (k/k_0)$ against p should be a straight line through the origin, of slope $-\Delta V^{\ddagger}/RT$. Such plots are used to evaluate numerically the volume of activation ($\Delta V^{\ddagger}$). Typical values range from +15 to -15 cm^3 mol^{-1}.

Experimental studies of the effect of pressure show that reactions can be divided into three categories in which:

(a) The rate of reaction is increased by the application of pressure (i.e. $\Delta V^{\ddagger}$ is negative). This behaviour is shown by reactions in which there is an increase in charge on forming the transition state, e.g. unimolecular (S_N1) solvolyses and reactions of alkyl halides with amines.

(b) Pressure has little effect and so $\Delta V^{\ddagger}$ is small. This behaviour is shown by reactions in which the transition state has the same charge as the products, e.g. reaction of hydroxide ions with primary alkyl halides (S_N2).

(c) The reaction is retarded by the application of pressure (i.e. $\Delta V^{\ddagger}$ is positive). This occurs in reactions between oppositely charged ions in which there is a decrease in charge on forming the transition state. An example is the nucleophilic substitution reaction

$$Co(NH_3)_5 Br^{2+} + OH^- \rightarrow Co(NH_3)_5 OH^{2+} + Br^-$$

Le Chatelier's principle suggests that reactions will be retarded by the application of pressure if the volume of activation is positive. Conversely, they will be accelerated by pressure if the volume of activation is negative.

To account for the magnitude and sign of the volume of activation it is necessary to introduce a new concept called *electrostriction*. This term describes the volume change in the solution caused by the orientation of polar solvent molecules around an ion in the solution. An analogy may help to visualize this effect.

Suppose a full box of matches is tipped in a pile on a table and then an attempt is made to refill the matchbox. Unless the matches are carefully oriented so that they are all parallel, they cannot be packed tightly, and the volume they occupy is greater than that of the matchbox.

If an ion is dissolved in a polar solvent, *the electrostatic field on the ion causes orientation of the dipoles* in the solvent molecules, so that they are arranged in a more ordered and hence more closely packed way. Consequently, the formation of a charge in solution decreases the total volume. The orientation of solvent molecules due to a dissolved ion is known as *electrostriction*. Categories (a), (b) and (c), listed above, can be explained in terms of charge type if it is assumed that electrostriction is the dominant effect. If charge is created in forming the transition state, there will be an increase in electrostriction and a decrease in volume. Such reactions are accelerated by increasing the pressure. Conversely, if charge is destroyed or distributed in forming the transition state, there will be a decrease in electrostriction and an increase in volume. Such reactions are retarded by an increase in pressure.

In the study of thermodynamics, the concept of entropy is associated with order–disorder changes. Since electrostriction is an effect of this type, it might be expected that the entropy change on forming the transition state, i.e. $\Delta S^{\ddagger}$ as defined in equation 5.19, should be related to electrostriction. It has in fact been found that there is a general correlation between entropy of activation and volume of activation.

As explained on page 60, a negative value of $\Delta S^{\ddagger}$ means that the rate of reaction is less than that obtained by simple collision theory. Hence reactions in category (a) which have negative values of both $\Delta V^{\ddagger}$ and $\Delta S^{\ddagger}$ are called *slow reactions*. In category (b) both $\Delta V^{\ddagger}$ and $\Delta S^{\ddagger}$ are small and these reactions are called *normal*. Finally, in category (c) both $\Delta V^{\ddagger}$ and $\Delta S^{\ddagger}$ are positive, and the reactions are said to be *fast*.

It must be emphasized that the terms *slow*, *normal* and *fast* are used in this context relative to simple collision theory, and that they have no absolute significance.

5.11 Effect of Ionic Strength on Reaction Rates

The transition state theory enables the effect of inert salts on reaction rates to be calculated. As in the previous section on the effect of pressure, the main idea derives from this theory. In equation 5.14 it is seen that *the rate constant of a reaction is proportional to a thermodynamic equilibrium constant*. The full expression for the thermodynamic equilibrium constant involves both concentration terms and activity coefficients. The latter are directly affected by the ionic strength of the solution, and so the position of equilibrium, and hence the concentration of the transition state, is altered by addition of salts. In dilute solution the activity coefficients can be calculated from the Debye–Hückel limiting law, and so the effect of inert salts on reaction rates can be predicted quantitatively.

It is first necessary to distinguish between the *primary* and the *secondary salt effects*. If reaction occurs directly between two ions then the effect of inert salts on the reaction rate is called the *primary salt effect*. However, in many cases of reactions between an ion and a molecule, pre-equilibria are involved. Consider for example the acid hydrolysis of sucrose in a buffer solution containing acetic acid and acetate ions. The hydrogen ions involved in the acid hydrolysis are also in equilibrium with the acetic acid and acetate ions. The effect of inert salts on such a reaction would be an example of the *secondary salt effect*. In this book, only the primary salt effect will be considered.

Suppose the reaction is similar to equation 5.10, namely,

$$A + B \rightleftharpoons C^{\ddagger} \rightarrow products$$

where A and B are the reactants and $C^{\ddagger}$ represents the transition state. Then

$$K^{\ddagger} = \frac{a_{C}{}^{\ddagger}}{a_{A} a_{B}} = \frac{(c/c^{\ominus})^{\ddagger}}{(a/a^{\ominus})(b/b^{\ominus})} \frac{y_{C}{}^{\ddagger}}{y_{A} y_{B}} \tag{5.27}$$

where the a terms are activities and the y terms are activity coefficients. The standard states* are set at unit concentration.

According to transition state theory from equation 5.14

$$k = mc^{\ddagger}/ab$$

Combining 5.14 and 5.27

$$k = mK^{\ddagger}y_{A}y_{B}/y_{C}{}^{\ddagger} \tag{5.28}$$

In dilute solutions, let the value of $k = k_{0}$. Under these conditions the activity coefficients term $y_{A}y_{B}/y_{C}{}^{\ddagger}$ will be approaching unity and so

$$k_{0} = mK^{\ddagger} \tag{5.29}$$

Combining 5.28 and 5.29 gives

$$k = k_{0}y_{A}y_{B}/y_{C}{}^{\ddagger} \tag{5.30}$$

---

*For reasons which are not discussed here, the standard states which are often assigned apply to hypothetical (ideal) solutions. These idealized standard states are particularly useful for electrolyte solutions. For further details consult: Robinson, R.A. and Stokes, R.M., *Electrolyte Solutions*, Second Edition, Butterworths (Revised 1968)

This result is known as the *Brönsted–Bjerrum equation*. Taking common logarithms of 5.30 and using M14

$$\log(k/k_0) = \log y_A + \log y_B - \log y_C^{\ddagger} \tag{5.31}$$

According to the Debye–Hückel limiting law

$$\log y = -Az^2\mu^{1/2} \tag{5.32}$$

where y is the activity coefficient of an ion in a dilute solution of ionic strength μ, z is the charge on the ion and A is a constant, numerically equal to 0.509 in aqueous solutions at 25 °C. The ionic strength is equal to the concentration for an electrolyte containing only univalent ions.

Using 5.32 and 5.31

$$\log_{10}(k/k_0) = -A\mu^{1/2}(z_A^2 + z_B^2 - z_C^2) \tag{5.33}$$

Now the charge on the transition state (z_C) must be the sum of the charges $(z_A$ and $z_B)$ on the reactants, i.e.

$$z_C = z_A + z_B \tag{5.34}$$

Putting 5.34 in 5.33 gives

$$\boxed{\log_{10}(k/k_0) = 2A\, z_A z_B \mu^{1/2}} \tag{5.35}$$

Thus the primary salt effect predicts that the plot of log k *against* $\mu^{1/2}$ *will be a straight line of slope* $2Az_A z_B$. This has been observed in many cases. For example, in aqueous solution the conversion of ammonium cyanate to urea

$$NH_4CNO \rightarrow CO(NH_2)_2$$

takes place. (This was the first synthesis of organic matter.) A straight line plot of log k against the square root of the molar concentration is obtained whose slope at 25 °C is -1.02. Since the slope is $2Az_A z_B$ and $A = 0.509$, the conclusion is that for this reaction $z_A z_B = -1$. This suggests that the reaction is a bimolecular one between the ammonium ion $(z_A = +1)$ and the cyanate ion $(z_B = -1)$.

The primary salt effect can, therefore, be used to deduce values of the charges on the reacting species in certain cases. It should be noted that if one of the reactants is uncharged, $z_A z_B$ is necessarily zero. There is, therefore, no primary salt effect in the reaction of an ion with a neutral molecule.

6

UNIMOLECULAR REACTIONS

6.1 Problem of First-order Reactions

First-order reactions present a special problem in kinetic studies. As mentioned previously, most reactions with a one-stage mechanism obey second-order rate laws since the rate of reaction is proportional to the number of binary collisions, which in turn is proportional to the product of two concentration terms or to the square of one concentration term. As an example, the homogeneous thermal decomposition of nitrous oxide in the gas phase

$$2N_2O \rightarrow 2N_2 + O_2$$

is a second-order reaction, as expected.

There are, however, many reactions in both the gas and the liquid phase which obey first-order kinetics. This poses the problem of why the reaction is not of second order. The answer must be found in the conditions or mechanism of the reaction.

Another way of realizing the difficulty raised by first-order reactions is to ask the question 'what is the origin of the energy of activation?' If a reaction is strictly first order then continually decreasing the concentration will not alter its half-life (equation 2.11). In the extreme case, an isolated molecule would still be expected to react with the same half-life as that observed at high concentrations, but it is difficult indeed to see how an isolated molecule can acquire the necessary energy of activation.

To answer these queries, the following examples will be considered in which first-order kinetics are observed.

6.2 Radioactive Decay

These are nuclear, not chemical, reactions, their rates being determined by changes in the nucleus rather than by changes in the energy levels of the outer electrons. They do not, therefore, come under the heading of chemical kinetics, but it is worth noting that the rates of radioactive decay accurately obey first-order laws. Temperatures of up to a few thousand degrees have no effect on the rate of nuclear decay as the activation energy comes from within the nucleus, and is, in any case, millions of times greater than the energies involved in molecular collisions. Since they are always first-order processes, rates of radioactive decay can be described by half-lives (equation 2.16).

6.3 Some Reactions in Solution

If a reaction occurs in a series of steps and the rate-determining step is unimolecular, then first-order kinetics will be observed. This occurs in the hydrolysis of t-pentyl iodide and is discussed in more detail in Section 7.5.

First-order kinetics are also often observed with bimolecular reactions in solution if one of the reactants is present in excess, as, for example, if the solvent is a reactant. Thus in the reaction

$$CH_3COCl + C_2H_5OH \rightarrow CH_3CO_2C_2H_5 + HCl$$

the concentration of ethanol remains effectively constant. Such a reaction is referred to as a *pseudo-first-order reaction.*

6.4 First-order Unimolecular Gas Reactions

First-order kinetics can be observed in the gas phase with homogeneous, heterogeneous and chain reactions. There are complex homogeneous gas reactions, such as the decomposition of dinitrogen pentoxide, which obey first-order rate laws. This particular case is discussed in Section 7.6, and heterogeneous and chain reactions are considered in Chapters 8 and 11, respectively. The remainder of this chapter will, therefore, be devoted to the theories which are used to interpret unimolecular gas reactions, i.e. reactions in which the activated complex is formed from a single reactant molecule. At normal pressures these are first-order reactions.

A reaction in the gas phase is assumed to be unimolecular if it (a) obeys accurately the first-order rate law, (b) is homogeneous, (c) is not a chain reaction, and (d) changes order from one to two when the pressure is reduced to a few millimetres of mercury. The absence of surface effects is shown if altering the ratio of surface area to volume (e.g. by adding glass beads or powdered glass) produces no change in the rate of reaction. The absence of chain reactions is shown by the use of inhibitors (see Chapter 11).

Many isomerization reactions in the gas phase are unimolecular, e.g.

$$\underset{\substack{\text{cyclopropane}}}{\overset{\text{CH}_2}{\underset{\text{CH}_2-\text{CH}_2}{\diagup\diagdown}}} \rightarrow \underset{\text{propene}}{CH_3CH=CH_2}$$

$$\underset{\substack{HCCO_2CH_3 \\ \| \\ HCCO_2CH_3 \\ \text{dimethyl maleate}}}{} \longrightarrow \underset{\substack{HCCO_2CH_3 \\ \| \\ CH_3CO_2CH \\ \text{dimethyl fumarate}}}{}$$

Often the first stage of complex multistage reactions is a unimolecular process, for example

$$\underset{\text{ethane}}{C_2H_6} \longrightarrow \underset{\text{methyl radicals}}{2CH_3\cdot}$$

6.5 The Lindemann Theory

The problem of the origin of the energy of activation in a unimolecular gas reaction is a difficult one. It was finally solved in 1923 by F.A. Lindemann. He pointed out that it is possible for molecules to receive their energy of activation by bimolecular collision, and still obey first-order kinetics. This can happen if there is a time delay between activation and reaction during which most of the energized molecules are deactivated by collision with normal molecules. If A represents the normal molecule and A* is the energized molecule, the *Lindemann mechanism* may be formulated as follows:

$$A + A \underset{k_{-1}}{\overset{k_1}{\rightleftharpoons}} A + A^* \tag{6.1}$$

$$A^* \overset{k_2}{\rightarrow} \text{products} \tag{6.2}$$

where k_1, k_{-1}, k_2 are the rate constants in 6.1 and 6.2.

The exact solution of the kinetic equations that describe this mechanism is difficult, but a simple result may be obtained using an approximation known as the *Stationary State Hypothesis*. This postulates that when reaction is brought about by very reactive molecules present at low concentrations, the concentration of such molecules may be regarded as constant.

If only a small fraction of the energized molecules (A*) reacts to give product (the majority being deactivated), a stationary concentration of A^* will be built up so that the rate of formation of A^* equals the rate of its removal. Under these circumstances the concentration of A^*, and hence the rate of reaction, will be proportional to the concentration of A. The reaction will therefore be first order. This argument can be seen mathematically by referring to the Lindemann mechanism shown in 6.1 and 6.2. By applying the stationary state hypothesis, then

$$\text{rate of formation of } A^* = \text{rate of destruction of } A^* \quad (6.3)$$

If a and a^* represent the concentrations of A and A^* at time t the stationary state hypothesis may be expressed by the equation

$$\frac{da^*}{dt} = 0 \quad (6.4)$$

which is equivalent to 6.3.

From equation 6.1

$$\text{rate of formation of } A^* = k_1 a^2$$

From 6.1 and 6.2

$$\text{rate of destruction of } A^* = k_{-1}a^*a + k_2a^*$$

Using 6.3

$$k_1 a^2 = k_{-1}a^*a + k_2a^*$$

Rearranging

$$a^* = \frac{k_1 a^2}{k_{-1}a + k_2} \quad (6.5)$$

From 6.2 the rate of formation of product is $k_2 a^*$. This is also the rate of reaction $(-da/dt)$. Hence from 6.5

$$\boxed{\frac{-da}{dt} = \frac{k_1 k_2 a^2}{k_{-1}a + k_2}} \quad (6.6)$$

At high pressures $k_{-1}a$ is much greater than k_2 (i.e. more molecules are deactivated than react). Hence k_2 can be ignored in comparison with $k_{-1}a$ and so 6.6 becomes

$$\frac{-da}{dt} = \frac{k_1 k_2 a}{k_{-1}} = ka \qquad (6.7)$$

Since $k_1 k_2/k_{-1}$ is itself a constant (k), equation 6.7 is a first-order law.

The above argument is based on the assumption that many more molecules of A^* are deactivated than react to give product. If the pressure of the gas is continuously reduced, the time between collisions increases, until it eventually becomes greater than the time interval between the energizing collision and reaction. As a result, the fraction of energized molecules that react gradually increases, until, at very low pressures, all the energized molecules react. When this happens the rate of reaction is proportional to the rate of formation of A^*, i.e. to $k_1 a^2$. At low pressures the reaction will therefore be of second order. This may be seen directly from equation 6.6. When a is sufficiently small, $k_{-1}a$ becomes smaller than k_2, when 6.6 becomes

$$\frac{-da}{dt} = k_1 a^2$$

which is a second-order equation.

Thus, below a certain pressure, usually 5–50 mmHg (about 0.7–7 kPa), the order of the reaction should change gradually from one to two. This effect is in fact observed with unimolecular reactions.

The Lindemann mechanism is supported by the effect of added inert gas. If a unimolecular gas reaction is carried out in the presence of hydrogen, with the reacting gas at a partial pressure that would correspond to a second-order reaction in the pure gas, it is found that the reaction is first order with the same rate constant as observed in the high-pressure gas reaction. This is to be expected from the fact that the essential feature of this mechanism is that the energized molecules can be deactivated by collision during the time delay between activation and reaction. This is not a chemical process and can be brought about by any molecule that collides with the energized molecule A^*. In the presence of hydrogen, the deactivating reaction

$$A^* + H_2 \rightarrow A + H_2$$

occurs rapidly compared with the conversion of A* to products, thus maintaining the basic requirements of the Lindemann mechanism.

Evidence of the existence of time lags in intramolecular processes comes from the phenomenon known as *predissociation* observed in molecular spectroscopy. In these spectra the vibrational bands for the activated molecules are present but the fine structure due to the rotation of the molecule is absent. The explanation given of this effect is that the electronically excited molecule exists for only a short time (about 10^{-12} s) after which it reacts chemically. This time is long enough for many vibrations to occur (a vibrational period is about 10^{-13} s), but is not long enough for the relatively slow process of rotation, and so there is no rotational fine structure.

6.6 The Hinshelwood Theory

The application of the collision theory of reaction rates to unimolecular reactions leads to a surprising result. If the majority of the molecules are deactivated before they react it would be expected that the P factor in equation 5.10 $[k = PZ \exp(-E^{\ddagger}/RT)]$ should be much less than unity. In practice it usually has a value of 10^3-10^4.

The reason for this is that the Boltzmann factor, $\exp(-E^{\ddagger}/RT)$, used to calculate the fraction of activated molecules (see equation 4.16) does not allow for the vibrational degrees of freedom in a molecule. It was shown by Hinshelwood that if the energy of activation $(E^{\ddagger})$ is distributed among several vibrational degrees of freedom, the fraction of molecules having the activation energy is

$$\frac{(E^{\ddagger}/RT)^{s-1}}{(s-1)!} \exp\left(\frac{-E^{\ddagger}}{RT}\right) \qquad (6.8)$$

where s = number of vibrational degrees of freedom. In a polyatomic molecule, therefore, s is large.

If $s = 1$, equation 6.8 reduces to

$$\frac{(E^{\ddagger}/RT)^0}{0!} \exp\left(\frac{-E^{\ddagger}}{RT}\right)$$

Using M6 and M7, $x^0 = 0! = 1$, i.e. 6.8 reduces to the normal form for the Boltzmann factor, namely $\exp(-E^{\ddagger}/RT)$.

The expression 6.8 becomes considerably larger than $\exp(-E^{\ddagger}/RT)$ when s is greater than 1. Thus if $E^{\ddagger} = 80\,000$ J mol^{-1} and $s = 8$ then at 298 K

$$\frac{(E^{\ddagger}/RT)^{s-1}}{(s-1)!} = \frac{1}{7!}\left(\frac{80\,000}{8.31 \times 298}\right)^7 = 9.36 \times 10^6$$

The value of s which will account for the experimental results is usually less than the number of vibrational degrees of freedom. This is probably due to a limited number of vibrational degrees of freedom being involved in the formation of the transition complex.

Unimolecular reactions are those in which there is a delay between the initial activation and the formation of the transition state. In such reactions it is necessary to distinguish between an *energized* and an *activated* molecule.

A molecule is said to be *energized* when it possesses the activation energy, although this energy may be distributed in various parts of the molecule. When this energy is localized in the part of the molecule undergoing reaction, then the molecule has become *activated*. It is the conversion of energized into activated molecules that causes the time delay characteristic of unimolecular reactions.

The Hinshelwood treatment assumes that the delay is due to the transfer of energy to the bond involved in the reaction. The theory also allows for the fact that the greater the energy in an energized molecule, the more rapidly it is converted into an activated molecule.

6.7 The RRK and the Slater Theories

Briefly, these theories seek to explain how energized molecules undergo conversion into activated molecules. The Rice, Ramsberger and Kassel treatment (known as the RRK theory) assumes that energy can flow freely between the various vibrational modes during each vibration of the energized molecule. Thus the conversion rate is of similar magnitude to the vibration frequency of the energized molecule.

According to the Slater theory, energy is not transferred freely amongst the various modes of vibration in the molecule. Instead, reaction can occur when certain molecular parameters, such as bond length, acquire a critical value. This can happen when particular modes of vibration in the molecule come suitably into phase.

Marcus modified the RRK model (the RRKM theory) to make allowances for factors such as zero-point energies and molecular rotation. For detailed discussion of these theories the reader should consult the specialist literature*.

The experimental evidence does not allow a clear-cut choice to be made between these theories in all cases. The mathematical analysis of unimolecular reactions is one of the more difficult topics in chemical kinetics, which cannot be dealt with here.

For several years the decomposition of dinitrogen pentoxide was the only known case of a first-order gas reaction which was homogeneous and not a chain reaction. It does not fit the Lindemann mechanism since it remains first order down to a pressure of 0.05 mmHg (about 7 Pa). This decomposition is discussed as a special case in Section 7.6.

*BUNKER, D.L., *Theory of Elementary Gas Reaction Rates*, Pergamon Press, Oxford (1966)

FORST, W., *Theory of Unimolecular Reactions*, Academic, London and New York (1973)

ROBINSON, P.J. and HOLBROOK, K.A., *Unimolecular Reactions*, Wiley, New York and Chichester (1972)

7

INVESTIGATION OF MECHANISM BY KINETIC METHODS

The kinetic method is a valuable tool in the investigation of reaction mechanisms. This is particularly so where the mechanism involves two or more successive reactions, for by kinetic measurements information can be obtained about the number and type of molecules involved in the slowest, or rate-determining, step. This information usually cannot be obtained by other methods.

To illustrate the applications of the kinetic method, several examples have been chosen in which kinetic studies have played a vital part in elucidating reaction mechanisms.

7.1 The Reaction of Acetone with Iodine

This reaction was studied by Lapworth in 1904 and was one of the first kinetic investigations of mechanism. If a dilute solution of iodine in acidic aqueous acetone is allowed to stand at room temperature, a slow reaction occurs represented by the equation

$$CH_3COCH_3 + I_2 \xrightarrow[\text{acid}]{\text{aqueous}} CH_3COCH_2I + H^+ + I^- \qquad (7.1)$$

The progress of the reaction may be followed by removing aliquots at selected intervals, adding excess of potassium iodide solution and titrating the iodine remaining with sodium thiosulphate solution. If the concentration of the acid catalyst is considerably greater than that of the iodine solution, the reaction is of zero order, i.e. the rate of reaction is constant with time and does not depend upon the concentration of iodine or acetone. *Figure 7.1* shows the plots of ideal results obtained from four experiments (a), (b), (c) and (d). The corresponding initial concentration of iodine increases steadily from experiment (a) to experiment (d). Of the four chemical species present, acetone, water and the acid catalyst are present in excess, and so their concentrations would not be expected to occur in the rate law. It is, however, most surprising that the rate of reaction does not depend on the concentration of the iodine.

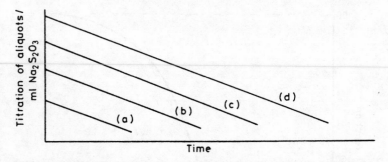

Figure 7.1   Reaction of iodine with acidified aqueous acetone

The fundamental principle used in interpreting this fact is that the kinetic results describe what is happening in the rate-determining step of the reaction. Hence the logical conclusion is that the rate-determining step does not involve iodine.

Having established this, it is now necessary to postulate a mechanism which is consistent with general chemical knowledge and with the observed kinetics. It is well known that some ketones (e.g. ethyl acetoacetate) undergo a slow reversible change from a keto to an enol form. If this type of equilibrium is set up in acetone the mechanism of iodination can be explained as

$$CH_3-\underset{\underset{O}{\parallel}\ keto}{C}-CH_3 + H^+ \xrightarrow[\text{controlling}]{\text{rate}} CH_3-\underset{\underset{OH}{\mid}\ enol}{C}=CH_2 + H^+ \qquad (7.2)$$

$$CH_3-\underset{\underset{OH}{\mid}}{C}=CH_2 + I_2 \longrightarrow CH_3-\underset{\underset{O}{\parallel}}{C}-CH_2I + H^+ + I^- \qquad (7.3)$$

where 7.2 is rate-controlling and imposes its limiting rate upon the more reactive iodination stage 7.3.

The acid-catalysed enol-forming step 7.2 consists of a proton transfer from H_3O^+ to carbonyl oxygen and abstraction of a proton from a methyl hydrogen by a neighbouring water molecule. This could be represented by the formation of an intermediate complex

$$\left[\begin{matrix} H & CH_3 & H \\ | & | & | \\ H-O\cdots H\cdots O\dot{=}C\dot{=}CH_2\cdots H\cdots O-H \end{matrix} \right]^+$$

In this mechanism the rate-determining step 7.2 does not involve iodine, in agreement with experiment. It may also be

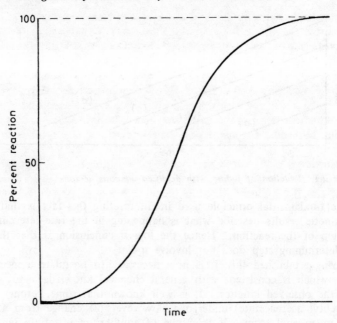

Figure 7.2   Autocatalytic reaction

predicted that the rate of bromination of acetone, and the rate of deuterium exchange, should both be equal to the rate of iodination, as both involve reaction with the highly reactive double bond in the enol form. This has been verified experimentally. Further confirmation of this mechanism has been obtained from studies with deuterated molecules, where the observed kinetic isotope effect (see p. 91) suggests that the removal of the methyl proton is the rate-limiting factor.

Another interesting aspect of this reaction is that the iodination step 7.3 generates acid. Hence the zero-order law will only be obeyed if the initial concentration of acid catalyst is relatively large, so that its concentration during the reaction remains effectively constant.

If the reaction is carried out with a small initial amount of acid catalyst the phenomenon of *autocatalysis* is observed. This is shown in *Figure 7.2*. The essential feature is that initially the rate of reaction increases with time as more acid is produced. By measuring the rate of the autocatalytic reaction in different proportions of acetone to water it has been shown that

$$\text{rate} \propto [\text{acetone}] \, [\text{acid}]$$

It may be concluded, therefore, that the reaction is second order, and probably bimolecular, as one molecule of acid and one molecule of acetone are involved in the rate-determining step.

7.2 Nitration of Aromatic Hydrocarbons

The kinetic studies of nitration are too extensive to discuss as a whole, but some particular points of interest will be mentioned here.

Benford and Ingold showed in 1938 that if nitration of aromatic hydrocarbons is carried out in organic solvents such as glacial acetic acid, and if an excess of nitric acid is used and the hydrocarbon is a reactive one, then the reaction is of zero order. Thus the same rate of nitration is measured for benzene, toluene and ethylbenzene in glacial acetic acid and in all three cases the rate is independent of the concentration of the hydrocarbon:

$$ArH + HNO_3 \rightarrow ArNO_2 + H_2O \qquad (7.4)$$

where ArH represents the aromatic compound.

Since the nitric acid is present in excess, its concentration would not be expected in the rate law. It is, however, surprising that the rate should be independent of the concentration of the hydrocarbon. The explanation of this fact is that the hydrocarbon cannot be involved in the rate-determining step of the reaction. It is therefore concluded that the slow step in the reaction is the conversion of nitric acid into the nitrating agent.

By combining spectroscopic, cryoscopic and kinetic evidence it was deduced that nitration is brought about by the nitronium ion (NO_2^+) which is formed slowly from nitric acid according to the equation

$$2HNO_3 \rightarrow NO_2^+ + NO_3^- + H_2O \qquad (7.5)$$

If the aromatic hydrocarbon has a high reactivity it removes nitronium ions as fast as they are formed, and the rate of reaction is determined by the reaction 7.5. The rate is therefore independent of the hydrocarbon concentration.

A different rate law is obtained when nitric acid replaces glacial acetic acid as the solvent. The process is then first order in the hydrocarbon ArH. Thus the rate-controlling stage is no longer the formation of nitronium ions (as in equation 7.5). Additional evidence about the rate-limiting stage which applies under this more acidic condition comes from the study of the kinetic isotope effect (see p. 91). Under the same experimental conditions the rates of nitration of benzene and hexadeuterobenzene are identical, i.e. there is no kinetic isotope effect. The interpretation of this result is that stretching of the carbon–hydrogen bond does not occur in forming the transition state. The rate-determining stage in this nitration is therefore believed to be the addition of the nitronium ion to the delocalized electrons in the aromatic ring, as shown in 7.6 and 7.7:

$$\text{ArH} + \text{NO}_2^+ \xrightarrow[\text{controlling}]{\text{rate}} [\text{ArH} \cdot \text{NO}_2]^+ \qquad (7.6)$$

$$[\text{ArH} \cdot \text{NO}_2]^+ \longrightarrow \text{ArNO}_2 + \text{H}^+ \qquad (7.7)$$

7.3 The Reaction of Hydrogen Peroxide with Iodide Ion

Hydrogen peroxide and iodide ion react in dilute aqueous acid at room temperature to form iodine according to the equation

$$\text{H}_2\text{O}_2 + 2\text{H}^+ + 2\text{I}^- \rightarrow 2\text{H}_2\text{O} + \text{I}_2 \qquad (7.8)$$

The reaction is rather fast and is usually followed by an indirect titration technique. A small known excess of sodium thiosulphate solution and some starch indicator are added to the acidified reaction mixture which is thoroughly stirred. The time for the formation of the blue starch–iodine colour is noted. Further thiosulphate is immediately added and the time for the reappearance of the blue colour is again noted. This procedure is repeated several times.

Iodine reacts rapidly and irreversibly with sodium thiosulphate according to the equation

$$\text{I}_2 + 2\text{S}_2\text{O}_3^{2-} \rightarrow 2\text{I}^- + \text{S}_4\text{O}_6^{2-} \qquad (7.9)$$

The blue complex is not formed until the thiosulphate has been consumed, and so in effect the reaction mixture titrates itself! The amount of hydrogen peroxide consumed at the time when the blue colour forms is equivalent to the volume of thiosulphate added.

An important kinetic consequence of equation 7.9 is that the concentration of iodide ions remains constant, for any iodine formed is immediately converted back to iodide by the thiosulphate. The observed kinetics will, therefore, depend only on the hydrogen peroxide concentration, and the results obtained show a first-order rate law as expected. On repeating the experiment with a different initial concentration of iodide, a first-order constant is again obtained, but it has a different numerical value. By carrying out several kinetic measurements it is found that the first-order rate constant is directly proportional to the initial concentration of iodide ions. Hence

$$\text{rate} \propto [\text{I}^-][\text{H}_2\text{O}_2] \tag{7.10}$$

The kinetic implications of equation 7.10 are that one iodide ion and one molecule of H_2O_2 are involved in the rate-determining step of the reaction. Since the mechanism of the reaction must also allow for the overall stoichiometry, the reaction may proceed as follows:

$$\text{H}_2\text{O}_2 + \text{I}^- \xrightarrow[\text{controlling}]{\text{rate}} \text{OI}^- + \text{H}_2\text{O} \tag{7.11}$$

followed by the rapidly established

$$\text{H}^+ + \text{OI}^- \rightleftharpoons \text{HOI} \tag{7.12}$$

and

$$\text{H}^+ + \text{I}^- \rightleftharpoons \text{HI} \tag{7.13}$$

then

$$\text{HOI} + \text{HI} \rightarrow \text{H}_2\text{O} + \text{I}_2 \tag{7.14}$$

The sum of equations 7.11–7.14 gives

$$\text{H}_2\text{O}_2 + 2\text{HI} \rightarrow 2\text{H}_2\text{O} + \text{I}_2$$

which is the overall chemical reaction (see equation 7.8). The rate of this reaction is increased by the addition of mineral acids, and comparative experiments must be carried out in solutions having the same acidity. An alternative or parallel mechanism involving hydrogen ions becomes important in solutions having a high hydrogen ion concentration. The overall rate law is of the form

$$-\text{d}[\text{H}_2\text{O}_2]/\text{d}t = k[\text{H}_2\text{O}_2][\text{I}^-] + k'[\text{H}_2\text{O}_2][\text{I}^-][\text{H}^+]$$

The second term may be interpreted as either a termolecular step

$$H_2O_2 + H^+ + I^- \rightarrow HOI + H_2O \qquad (7.15)$$

or a bimolecular step

$$H_2O_2 + HI \rightarrow HOI + H_2O \qquad (7.16)$$

which follows the initial and rapidly established equilibrium 7.13.

It must be stressed that these mechanisms are interpretations of the kinetic rate laws. Other equally valid reaction routes might be written based on the observed rate laws. Additional information is required before a particular mechanism sequence is established as unique.

7.4   The Oxidation of Nitric Oxide

This reaction obeys the equation

$$2NO + O_2 \rightarrow 2NO_2 \qquad (7.17)$$

and since it involves a decrease in number of molecules, it may be followed manometrically by measuring the decrease in pressure at constant volume. The results show that the reaction obeys the third-order rate law

$$\frac{-d[NO]}{dt} = k[NO]^2[O_2] \qquad (7.18)$$

A singular feature of the reaction is that it has an apparently negative energy of activation; that is, that the rate of reaction decreases as the temperature increases.

Third-order reactions are relatively uncommon between stable molecules. Statistically it can be shown that the probability of a three-body activated collision occurring is very small compared with the probability of an activated two-body collision. Hence the mechanism is unlikely to be termolecular

The nitric oxide molecule (like nitrogen dioxide, NO_2) is unusual in that it contains an odd number of electrons. It is well known that nitrogen dioxide tends to dimerize according to the equation

$$2NO_2 \rightleftharpoons N_2O_4 \qquad (7.19)$$

Furthermore, the N_2O_4 is thermally unstable and as the temperature is increased the equilibrium is driven from right to left.

The kinetic results for the oxidation of nitric oxide can be explained if it is assumed that an equilibrium similar to 7.19 is set up, i.e.

$$2NO \rightleftharpoons N_2O_2 \qquad (7.20)$$

followed by

$$N_2O_2 + O_2 \xrightarrow[\text{controlling}]{\text{rate}} 2NO_2$$

The observed order of reaction can be explained if only the N_2O_2 molecules react with oxygen. If K is the equilibrium constant for 7.20

$$K = \frac{[N_2O_2]}{[NO]^2}$$

i.e.

$$[N_2O_2] = K[NO]^2 \qquad (7.21)$$

Assuming a second-order reaction between N_2O_2 and O_2, its rate is proportional to $[N_2O_2][O_2]$. Hence, using 7.21

$$\text{overall rate} = \frac{-d[NO]}{dt} = k[NO]^2[O_2]$$

where k is a third-order rate constant.

This is in agreement with the experimental result for the rate law 7.18. There is some direct experimental evidence for 7.20 since N_2O_2 has been isolated at low temperatures. Also, at room temperature there are some bands in the ultraviolet spectrum of nitric oxide whose intensities are proportional to the square of the pressure. Equation 7.21 shows that these could be attributed to N_2O_2.

An explanation of the negative energy of activation of the reaction is that the equilibrium 7.20 is affected by temperature in the same way as is the equilibrium 7.19. Although raising the temperature increases the rate constant for the reaction of N_2O_2 with oxygen, this is more than offset by the decrease in the concentration of N_2O_2 brought about by thermal decomposition.

One final piece of evidence in favour of this mechanism is that the reactions of nitric oxide with hydrogen, chlorine and bromine respectively are all third order, and can be explained in terms of an intermediate N_2O_2. The only examples of third-order gas reactions are those involving nitric oxide.

7.5 Nucleophilic Substitution of Alkyl Halides

Alkyl halides (RX) will, in certain cases, react with hydroxide ions in aqueous ethanol in such a way that the halide (X) is replaced by the hydroxyl group:

$$RX + OH^- \rightarrow ROH + X^- \qquad (7.22)$$

This substitution reaction is classified as nucleophilic since the reagent (in this case OH^-) is one that will attack a positive centre (i.e. a region of low electron density).

There are two distinct mechanisms by which a nucleophilic substitution reaction, such as 7.22, can occur. These are

(a) a one-step bimolecular process

$$RX + OH^- \rightarrow ROH + X^- \qquad (7.23)$$

(b) a two-step process, involving a unimolecular ionization

$$RX \xrightarrow[\text{controlling}]{\text{rate}} R^+ + X^-$$

$$R^+ + OH^- \rightarrow ROH$$

$$\left. \right\} \quad (7.24)$$

Reactions of both types have been extensively studied by Hughes and Ingold, who use the terms S_N2 and S_N1 to represent the mechanisms in 7.23 and 7.24, respectively. The term S_N1 is a convenient abbreviation for a unimolecular nucleophilic substitution; similarly S_N2 refers to bimolecular nucleophilic substitution. Kinetically, it is expected that the S_N1 reaction will have an order of one, but the S_N2 reaction will have an order of two, unless one of the reactants is present in excess when the reaction will then have an order of one.

The reaction of straight-chain primary alkyl halides with hydroxide ions in aqueous ethanol obeys the second-order rate law and is an S_N2 process. With tertiary alkyl halides, however, the strength of the carbon–halogen bond is weakened and ionization is easier. This is because the electron-releasing effect of a tertiary group is greater than that of a straight chain. Thus in the tertiary halides the carbon–halogen bond is polarized in the direction that facilitates ionization.

The reactions of t-pentyl and t-butyl iodides with dilute sodium hydroxide solution are found to be first order and the rate is independent of the concentration of hydroxide ions. The possibility of direct reaction with the solvent is ruled out, since the hydroxide ion is a much more powerful base than the solvent, and if it does not react directly with the halide, it is unlikely that the solvent will. The kinetics therefore suggest an S_N1 mechanism for these reactions.

A reaction with interesting kinetics is the hydrolysis of but-2-yl sulphate by hydroxide ions in aqueous solution:

$$ROSO_3^- + OH^- \rightarrow ROH + SO_4^{2-} \qquad (7.25)$$

where R is the group $(CH_3CH_2CHCH_3)$. This is a first-order reaction, the rate being independent of the concentration of hydroxide ions. This suggests that the mechanism should be S_N1. However, by using optically active but-2-yl sulphate and measuring the initial and final optical rotation, it is found that the resultant butan-2-ol suffers an inversion of configuration (Walden Inversion) but is still optically active. In the S_N1 mechanism an alkyl cation (R^+) having a planar structure is postulated. When the hydroxide ion attacks the planar structure (R^+), equal quantities of *dextro* and *laevo* forms are produced and the sample should lose its optical activity. This is contrary to the observed facts.

Examination of the equation of reaction 7.25 shows that the two reactants are both negatively charged ions, which will tend to repel each other electrostatically. As a result, the uncharged water molecule is a more powerful reagent than the hydroxide ion in this particular case and the reaction occurs according to the two-stage process

$$RSO_4^- + H_2O \xrightarrow[\text{controlling}]{\text{rate}} ROH + HSO_4^-$$

$$HSO_4^- + OH^- \rightarrow H_2O + SO_4^{2-}$$

The reaction is therefore S_N2 but since the water is present in large excess, the kinetics will obey a first-order rate law. The S_N2 mechanism will explain the optical activity results stated above.

The distinction between S_N1 and S_N2 reactions is of great importance in organic chemistry, where it has been applied to evaluate the effect of substituents on substitution reactions.

7.6 The Thermal Decomposition of Dinitrogen Pentoxide

This reaction is of historical importance as it was the first example to be discovered of a homogeneous first-order gas reaction (1921). The reaction can be conveniently followed manometrically since it is accompanied by an increase in pressure at constant volume

$$2N_2O_5 \rightarrow 4NO_2 + O_2 \qquad (7.26)$$

Careful examination has shown that the rate is not changed by surface effects and that the reaction is uncatalysed. However, it has one surprising feature. According to the Lindemann theory of unimolecular gas reactions (Section 6.5), the order should change from one to two as the pressure is lowered. Further, the critical pressure at which the kinetic order changes can be calculated and it should be several millimetres of mercury. However, the decomposition of dinitrogen pentoxide accurately obeys the first-order rate law down to a pressure of 0.05 mmHg (7 Pa) and eventually becomes second order at 0.004 mmHg (0.5 Pa). It is impossible to account quantitatively for these pressure limits in terms of a simple unimolecular decomposition.

Ogg (1947) proposed the following mechanism which is consistent with the known experimental facts:

Step 1. A rapidly established equilibrium

$$N_2O_5 \underset{k_{-1}}{\overset{k_1}{\rightleftharpoons}} NO_2 + NO_3 \qquad (7.27)$$

Step 2. A rate-controlling reaction

$$NO_2 + NO_3 \overset{k_2}{\rightarrow} NO_2 + O_2 + NO \qquad (7.28)$$

which is followed by

Step 3. $\qquad NO + NO_3 \rightarrow 2NO_2 \qquad (7.29)$

in which the nitric oxide molecule is removed from the reaction system virtually as soon as it is formed.

The NO_3 molecule is not one of the stable oxides of nitrogen. It has not been isolated in the pure state, but its existence has been claimed in the products of a glow discharge of a mixture of nitrogen dioxide and oxygen. It was postulated in this mechanism by Ogg in order to explain the kinetic results.

This mechanism is consistent with first-order kinetics in which the observed first-order rate constant (k) is given by

$$k = \frac{2k_1 k_2}{k_{-1} + 2k_2} \qquad (7.30)$$

This may be shown by applying the stationary state hypothesis (Section 6.5) to the unstable intermediate NO_3.

At time t, let a = the concentration of $N_2 O_5$
b = the concentration of NO_2
c = the concentration of NO_3

Using equations 7.27–7.29

$$\frac{dc}{dt} = k_1 a - k_{-1} bc - 2k_2 bc = 0 \qquad (7.31)$$

The factor of 2 is used because for each NO_3 destroyed in 7.28 one NO molecule is formed which rapidly destroys another NO_3 by reaction 7.29.

From equation 7.31

$$bc = \frac{k_1 a}{k_{-1} + 2k_2} \qquad (7.32)$$

Since step 1 (7.27) is a rapidly established equilibrium, the rate of reaction is governed by step 2 (7.28) and hence

$$\frac{-da}{dt} = 2k_2 bc \qquad (7.33)$$

Substituting 7.33 in 7.32

$$\frac{-da}{dt} = \frac{2k_1 k_2 a}{k_{-1} + 2k_2} \qquad (7.34)$$

Remembering that a represents the concentration of dinitrogen pentoxide, equation 7.34 states that the rate of reaction of dinitrogen pentoxide is proportional to its concentration, the proportionality constant (i.e. the rate constant) being $2k_1 k_2/(k_{-1} + 2k_2)$, in agreement with equation 7.30 and thus gives a satisfactory account of the first-order kinetics.

The postulated mechanism has been verified in the following ways:

(a) If 7.27 is a rapidly established equilibrium, there should be isotope exchange between dinitrogen pentoxide and nitrogen dioxide enriched in the isotope ^{15}N:

$$N_2O_5 + {}^{15}NO_2 \rightleftharpoons NO_2 + NO_3 + {}^{15}NO_2 \rightleftharpoons {}^{15}N_2O_5 + NO_2$$

$$(7.35)$$

This exchange has been demonstrated experimentally. Furthermore, the rate of isotope exchange is a measure of the rate of the unimolecular decomposition of N_2O_5 shown in 7.27. The value of k_1 obtained by this method is much greater than k (the observed first order rate constant for the whole process), which is in agreement with the postulate that 7.28 is the rate-determining step.

By measuring the rate of isotope exchange at various pressures it is possible to show that the rate of the unimolecular step 1 varies with pressure in accordance with the Lindemann theory. In fact k_1 begins to change from first to second order as pressure is lowered below 50 mmHg (7 kPa).

(b) When the reaction is carried out in the presence of added nitric oxide, the rate of reaction

$$NO + N_2O_5 \rightarrow 3NO_2$$

is independent of the concentration of nitric oxide. Under these circumstances, step 2 of the proposed mechanism is bypassed and the reversible decomposition in step 1 becomes rate controlling. The variation in order with pressure expected for a unimolecular reaction is again observed for the thermal decomposition of dinitrogen pentoxide in the presence of excess nitric oxide. This provides indirect kinetic evidence for the existence of the equilibrium postulated in step 1 of the Ogg mechanism.

7.7 The Reaction of Bromide and Bromate Ions in Acidic Aqueous Solution

The stoichiometric reaction

$$5Br^- + BrO_3^- + 6H^+ \rightarrow 3Br_2 + 3H_2O \qquad (7.36)$$

is frequently used in quantitative analysis. An aqueous bromide-bromate solution is quite stable in neutral solution, but on acidifying it with a strong acid, bromine is rapidly formed.

An unusual feature of the reaction is that it is of the fourth order, obeying the equation

$$\text{rate} = k[\text{Br}^-][\text{BrO}_3^-][\text{H}^+]^2 \tag{7.37}$$

The dependence of the rate on the square of the hydrogen ion concentration is also unusual. Thus in water at pH 7, $[\text{H}^+]$ = 10^{-7} mol l^{-1}, which is one millionth of the value in, say, 0.1 N perchloric acid. Consequently, the rate of reaction in water is $10^6 \times 10^6 = 10^{12}$ times as slow as in the acid. Putting this another way, if the half-life of reaction 7.36 is 0.1 s in the acid, then the half-life is about 3000 years in a neutral solution. This accounts for the stability of standard bromide—bromate solutions.

The reaction has been studied in the presence of added inert salts such as sodium perchlorate. The rate of reaction is greatly decreased by the presence of these salts, but in a more involved manner than is required by the simple theory of the primary or secondary kinetic salt effects (Section 5.11). Further, if the reaction is carried out in heavy water (deuterium oxide, D_2O), the rate is about three times as fast as it is in ordinary water. (This is the reverse kinetic isotope effect discussed in Section 7.9.)

The kinetic evidence is not sufficient to elucidate the mechanism of this complex multi-stage reaction, but an explanation of the fourth-order rate law can be given in terms of a bimolecular rate-determining stage between the species $HBrO_3$ and HBr. (The exact nature of these species in aqueous solution is not too clear, but they might be regarded as ion-pairs.)

It is assumed that both the acids dissociate according to the following equilibria

$$HBrO_3 \overset{K_1}{\rightleftharpoons} H^+ + BrO_3^- \tag{7.38}$$

$$HBr \overset{K_2}{\rightleftharpoons} H^+ + Br^- \tag{7.39}$$

where K_1 and K_2 are the dissociation constants for the reactions shown.

On this proposed mechanism

$$\text{rate} = k[HBr][HBrO_3] \tag{7.40}$$

Using the expressions for the equilibrium constants defined by 7.38 and 7.39, equation 7.40 becomes

$$\text{rate} = \frac{k[H^+][Br^-][H^+][BrO_3^-]}{K_2 K_1} = k'[Br^-][BrO_3^-][H^+]^2$$

(7.41)

where

$$k' = k/K_1 K_2$$

(7.42)

which is the experimental fourth-order rate constant for the bromide–bromate reaction.

This mechanism does account for the dependence of the reaction rate on the square of the hydrogen ion concentration. It can also account for the decrease in rate with ionic strength, since formal acid dissociation constants (i.e. those formulated in terms of concentrations rather than activities) increase when inert salts are added. In deriving equation 7.41 formal equilibrium constants were used, and so the fourth-order rate constant k' in 7.42 will decrease if the values of K_1 and K_2 are increased by adding inert sal

The effect observed in heavy water can be explained if it is remembered that D_2O is a poorer ionizing solvent than H_2O, and so K_1 and K_2 have smaller values in heavy water than in ordinary water. This point is illustrated by the fact that at 25 °C the pK value for the self-ionization of water is 14.01. The corresponding value for heavy water is 14.85, showing that heavy water is ionized to a lesser extent than ordinary water.

There are, however, several other mechanisms that can account for the fourth-order rate law and the salt and isotope effects in the qualitative manner used in the preceding paragraphs. In one possible mechanism, there is a rapid pre-equilibrium

$$H^+ + HBrO_3 \xrightleftharpoons{\text{fast}} H_2O + BrO_2^+$$

(7.43)

followed by the rate-determining step

$$BrO_2^+ + Br^- \xrightarrow{\text{slow}} \text{reaction intermediates}$$

This mechanism will fit the observed kinetic data since it involves two equilibria 7.38 and 7.43. Thus

$$\text{rate} \propto [BrO_2^+][Br^-] \propto [H^+][HBrO_3][Br^-]$$
$$\propto [H^+]^2[BrO_3^-][Br^-]$$

(7.44)

which is the required fourth-order rate law. The arguments used previously about the salt and isotope effects also apply in this case.

Although the kinetics of the bromide–bromate reaction were first studied at the end of the last century, the mechanism of this reaction sequence is not known with certainty. This example again illustrates the limitations of the kinetic method, which in certain cases cannot distinguish between various possible mechanisms. To do this, additional information is required, such as the positive identification of proposed intermediates.

7.8 The Kinetic Isotope Effect

Additional insight into reaction mechanisms is possible when kinetic studies are carried out with reactant molecules that have been modified by isotope substitution. It may happen that the rate constant of the reaction is altered by the isotopic substitution. Such a change in rate is referred to as the *kinetic isotope effect.*

The existence of this effect suggests that the bond abnormally stretched in the transition state involves one or more of the isotopically substituted atoms. Thus kinetic studies, as well as indicating the molecules in the rate-determining stage, can also point to the particular bonds involved.

Large changes in rate have been observed in many reactions following isotopic substitution of hydrogen by deuterium since the relative change in mass when this element is involved is very large. Indeed, the rate change is so pronounced that the use of deuterium-substituted compounds has been suggested for the slowing of unwanted reactions, such as the oxidative deterioration of engine lubricants.

With the increasing availability of deuterium and tritium (the isotopes of hydrogen of mass numbers 2 and 3, respectively), many compounds have been made in which a hydrogen atom has been substituted by one of these isotopes. Heavy water (deuterium oxide) is in fact only about ten times as dear as ethanol for human consumption on which tax has been paid! Tritium is radioactive and is a weak β emitter, with a half-life of 18 days. Tritium oxide costs (1980) about £30 per curie. Both isotopes are, therefore, readily available for kinetic work.

A detailed account of kinetic isotope effects requires a far-ranging discussion beyond the scope of this text. However, a semi-quantitative explanation can be attempted by considering the stretching and contracting of the (only) bond in a diatomic molecule. As the internuclear distance departs from the average bond

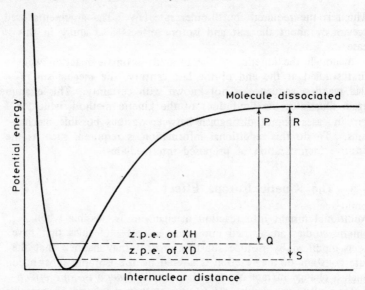

Figure 7.3 Effect of nuclear mass on dissociation energy

length, the potential energy of the molecule increases. Since the interatomic force at any internuclear distance is independent of nuclear mass, the potential energy curve shown in *Figure 7.3* applies equally to the unmodified and the isotopically substituted molecule.

However, the nuclear mass does affect the spacing between the discrete vibrational energy levels. In particular, nuclear mass has a relatively large effect on the level of the lowest vibrational energy state. This level is known as the *zero-point energy* of the molecule. Most molecules at ordinary temperatures are at the zero-point energy level.

For a molecule XH, the zero-point energy (E_0) is given by

$$E_0 = \frac{h}{4\pi} \left[\frac{k(m_X + m_H)}{m_X m_H} \right]^{\frac{1}{2}} \tag{7.45}$$

where h is Planck's constant, k is the force constant of the X–H bond and m_X and m_H are the masses of the atoms X and H. Usually the hydrogen atom is attached to a much more massive atom, so as a good approximation 7.45 becomes

$$E_0 = \frac{h}{4\pi} \left(\frac{k}{m_H} \right)^{\frac{1}{2}} \tag{7.46}$$

The zero-point energy is therefore inversely proportional to the square root of the mass of the lighter atom of the bond (in this case m_H).

Since the atomic masses of hydrogen, deuterium and tritium are in the ratio of 1 : 2 : 3, the corresponding zero-point energies will be in the ratio of

$$1 : 1/2^{1/2} : 1/3^{1/2}$$

$$(7.47)$$

i.e. $\quad\quad\quad 1 : 0.707 : 0.577$

The X–H, X–D and X–T bonds have the same potential energy curve, which is shown in *Figure 7.3*. The relative zero-point energies of the X–H and X–D bonds are also shown. The X–T zero-point energy lies below that of the X–D bond, as indicated by the ratios in equation 7.47.

The zero-point energies (abbreviated to z.p.e.) of the XH and XD molecules are shown by horizontal lines. The dissociation energy of XD (shown by the line RS) is clearly greater than that for the molecule XH (shown by the line PQ).

When the X–H bond is abnormally stretched in forming a transition state, a similar effect occurs (see *Figure 7.4*). The molecule containing the heavier isotope requires a larger activation energy, and so reacts at a slower rate in a given reaction.

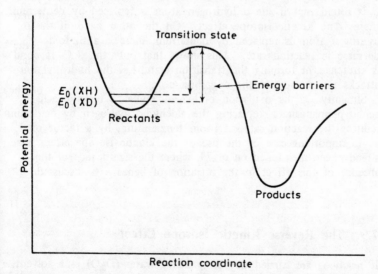

Figure 7.4 An illustration of XH and XD zero-point energies and corresponding energy barriers

An alternative way of visualizing the effect of zero-point energy on the rate of a reaction is to consider the effect of isotopic substitution on activation energies. This is illustrated in *Figure 7.4.* The reaction coordinate used on the horizontal axis is a measure of the extent to which a pair of molecules have completed their interaction.

With polyatomic molecules, a similar situation applies. Consider a typical saturated organic molecule in which a C–H bond is stretched until the H atom is transferred to some other molecule. The zero-point energy of C–H is about 18 kJ mol^{-1}. So, using equation 7.47 the comparable zero-point energy of C–D is about 13 kJ mol^{-1}. Thus the difference in activation energy for the C–H and C–D bonds can be as much as 5.5 kJ mol^{-1}, which at room temperature corresponds to a ten-fold decrease in rate for the reaction involving the C–D bond. On substituting tritium (mass number 3) for hydrogen, the rate can be decreased by a factor of about 25. These very large changes can easily be detected by kinetic experiments.

Some examples will show the use in mechanistic studies of the kinetic isotope effect. In the elimination reaction

$$C_6H_5CH_2CH_2Cl + OH^- \rightarrow$$

$$C_6H_5CH=CH_2 + H_2O + Cl^-$$

it is found that if the a hydrogen atom is replaced by deuterium, there is no kinetic isotope effect. On the other hand, if the β hydrogen atom is replaced by deuterium, there is a six-fold decrease in reaction rate. This shows that only the β C–H bond is stretched in forming the transition state, i.e. the hydroxyl ion attacks the β and not the a carbon atom.

Similarly, in the oxidation of benzaldehyde to benzoic acid by acidic permanganate, replacing the aldehydic hydrogen by deuterium reduces the reaction rate, at room temperature, by a factor of 7.

An important case of the use of the kinetic isotope effect has already been mentioned on p. 79, where the significance of the absence of this effect in the nitration of benzene is discussed.

7.9 The Reverse Kinetic Isotope Effect

If reactions are carried out using heavy water (D_2O) as a solvent, in some cases the reaction rate is greater than that in ordinary water. This is known as the *reverse kinetic isotope effect.* The acid-catalysed hydrolysis of aliphatic esters and the bromide-bromate reaction show this effect.

The simplest explanation is to regard D^+ as a stronger acid (in the Brönsted sense) than H^+. Hence reactions in which there is a rapid and reversible protonation as a first stage will be faster in D_2O, as a higher concentration of the intermediate will be built up.

8

HETEROGENEOUS CATALYSIS

8.1 Properties of a Catalyst

It is common experience that the rate of a reaction can be markedly increased by the presence of a small amount of another substance which is not used up in the reaction. Such a substance is called a *catalyst*. A well-known example of catalysis occurs in the laboratory preparation of oxygen by the thermal decomposition of potassium chlorate in the presence of manganese dioxide.

A catalyst has the following properties:

(a) It increases the rate of the reaction when present in small amounts.

(b) It is unchanged chemically at the end of the reaction, although it participates cyclically in the reaction and is therefore continuously regenerated. However, the *physical form* of a solid catalyst may be different at the end of a reaction.

(c) It can be specific to a particular reaction.

(d) It cannot alter the enthalpy change, free energy change or equilibrium constant of the reaction. Consequently, it must increase the rate of the reverse reaction as much as it increases the rate of the forward reaction.

There are some substances which can decrease the rate of reaction very markedly when present in trace amounts. They are sometimes called *negative catalysts*, but this is a misleading name as these substances may be consumed in the reaction. The term *inhibitor* or *catalyst poison* is preferred (see p. 105).

In many chemical reactions, including several of industrial importance, reaction occurs on a solid surface and not uniformly throughout the gas or liquid surrounding the solid. Such reactions are *heterogeneous*. When a surface is involved in the reaction mechanism, the rate of reaction at a given temperature depends on the pressure or concentration of the reagents and on the area and chemical nature of the surface. It is for this reason that solid catalysts are used in a finely divided form, so as to have a large surface area.

The most remarkable feature of heterogeneous reactions is the chemical specificity of the solid (catalyst) phase. Not only do different solids produce rates of reaction differing by many powers of ten, but they may also yield different products from the same initial compound. Two examples of this type of behaviour are shown below for the thermal decomposition of ethanol vapour and of formic acid vapour

$$C_2H_5OH \xrightarrow{\text{alumina}} C_2H_4 + H_2O$$

$$C_2H_5OH \xrightarrow{\text{copper}} CH_3CHO + H_2$$

$$HCO_2H \xrightarrow{\text{alumina}} H_2O + CO$$

$$HCO_2H \xrightarrow{\text{copper}} H_2 + CO_2$$

These examples demonstrate the fact that the catalyst cannot function merely by concentrating the reacting gas in the surface layer and so increasing the rate of reaction at the surface. The formation of different products with particular catalysts shows that the surface atoms of the solid are involved in the reaction mechanism. This seems to be anomalous, as it appears that the catalyst has altered the thermodynamic properties of the reaction, but this is not so.

In the quoted examples, the reactant can undergo alternative chemical changes. It will not necessarily follow the route with the greatest decrease in free energy. The reaction is controlled by the activation energies of the alternative reaction routes. Thus a specific catalyst lowers the activation energy of a particular reaction route and the increased rate of reaction results in a greater proportion of the faster-forming product.

The role of the solid surface in chemical reactions was explained by Langmuir in 1916. He suggested that the adsorbed molecules were held to catalytic surfaces by chemical bonds. This process is nowadays referred to as *chemisorption*. It contrasts with physical adsorption in which the molecules are held to the surface by weak van der Waals forces of attraction. Chemisorption is characterized by a high heat of adsorption (80–800 kJ mol^{-1}) which is comparable with the heat evolved in a chemical change. Indeed the heat of adsorption of the first fraction of oxygen on to charcoal is over twice the heat of combustion of carbon. This is because no energy has to be used to break bonds in the carbon lattice when forming a surface compound and so more heat is evolved than in combustion.

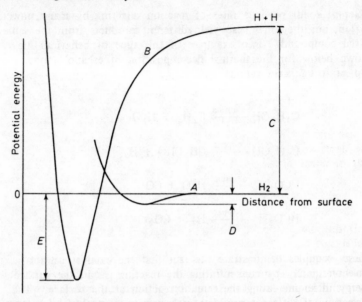

Figure 8.1 Potential energy diagrams for the adsorption of hydrogen on a nickel surface

Another striking feature of chemisorption is that it can occur at temperatures far above the boiling point of the adsorbed substance (e.g. ammonia on tungsten at 800 °C), which again points strongly to firm bonding between the surface and the gas.

8.2 Mechanism of Chemisorption

The mechanism by which reactant molecules are adsorbed on to a solid catalyst will be considered using as an illustration the potential energy changes for the adsorption of hydrogen on a nickel surface. A diagrammatic representation of the two types of adsorption is shown in *Figure 8.1* in which curve A shows the potential energy changes during the process of physical adsorption of hydrogen molecules

$$H_2 + Ni \text{ surface} \rightarrow H_2 \cdots Ni \text{ surface}$$

Curve B shows the potential energy changes of chemisorption direct from hydrogen atoms

$$2H + Ni \text{ surface} \rightarrow 2H-Ni \text{ surface}$$

The energy difference C represents the energy of the dissociation

$$H_2 \rightarrow 2H$$

The depth of the potential energy minimum D corresponds to the heat of physical adsorption

$$H_2 + Ni \text{ surface} \rightarrow H_2 \cdots Ni \text{ surface}$$

The depth of the potential energy minimum E corresponds to the heat of chemisorption

$$H_2 + Ni \text{ surface} \rightarrow 2H\text{–}Ni \text{ surface}$$

During chemisorption, chemical bonds are formed between the nickel surface and the hydrogen atoms. The large amount of energy required for the complete dissociation of the hydrogen molecules into hydrogen atoms is indicated, in *Figure 8.1*, by the energy difference C. The formation of chemical bonds between hydrogen atoms and the nickel surface is accompanied by a large decrease in potential energy, shown in curve B. The variation of potential energy with distance from the surface for the physical absorption of hydrogen molecules is shown in curve A.

As the hydrogen molecules approach the surface, the weak, (relatively) long-range van der Waals forces of attraction between the hydrogen molecules and the nickel surface bring about a decrease in potential energy. This is changed to a steep increase in potential energy as closer approach to the surface involves strong forces of repulsion. Physical adsorption is thus established at the equilibrium distance corresponding to the shallow potential energy minimum D.

By the process of physical adsorption, the hydrogen molecules can come close to the nickel surface without the large energy requirement for prior dissociation into hydrogen atoms. Then the hydrogen molecules only have to surmount the relatively low energy barrier corresponding to the point of intersection of curves A and B for chemisorption to be possible.

This mechanism can be thought of as a concerted process in which the stretching and breaking of the hydrogen–hydrogen bond is assisted by the incipient bonding of the hydrogen atoms to the nickel surface. By chemically interacting with the nickel atoms in the surface, the hydrogen atoms approach close to the short-range equilibrium distance of the deep potential energy minimum E.

The mechanism by which a gas reaction can be heterogeneously catalysed at a solid surface will often proceed by this type of route.

8.3 Langmuir Adsorption Isotherm

When the adsorbed gas is chemisorbed, there is a certain limiting amount of gas that can be held on the surface. Langmuir concluded that this limit is reached when the surface is covered by a layer one molecule thick.

Using the idea that there is a dynamic rather than a static equilibrium between the adsorbed molecules and the molecules present in the gas phase, it is possible to deduce a relationship between the fraction of the surface covered and the pressure of the gas at constant temperature.

Let p = pressure of the gas
     θ = fraction of surface covered by gas
then $(1 - \theta)$ = fraction of surface not covered.

If a dynamic equilibrium is set up then

$$\text{rate of evaporation}\ =\ \text{rate of adsorption} \qquad (8.1)$$

The rate of evaporation depends only on the amount of gas already adsorbed, which is proportional to θ. Using M1

$$\text{rate of evaporation}\ =\ k_1 \theta \qquad (8.2)$$

The rate of adsorption is proportional to the area of uncovered surface and to the pressure in the gas phase, since the pressure determines the number of collisions between gas molecules and solid surface per unit area. Hence

$$\text{rate of adsorption} = k_2(1 - \theta)p \qquad (8.3)$$

Using the fact that rate of evaporation equals rate of adsorption (equation 8.1), then

$$\boxed{k_1 \theta\ =\ k_2(1 - \theta)p} \qquad (8.4)$$

Equation 8.4, which refers to the adsorption of a single gas at a constant temperature, is known as the *Langmuir Adsorption Isotherm.* It may be used to deduce the order of simple heterogeneous reactions if it is assumed that the rate of reaction is proportional to the amount of adsorbed gas.

There are four special cases of particular kinetic importance, which can be interpreted using the ideas outlined above.

Case 1.   *The adsorption is slight*

In this case, θ is much less than unity and so $(1 - \theta)$ is nearly equal to unity. Equation 8.4 then becomes

$$k_1 \theta = k_2 p \tag{8.5}$$

i.e. the amount of adsorbed gas is proportional to the pressure of the gas.

Since the rate of reaction depends on the amount of adsorbed gas, equation 8.5 shows that the rate of reaction will be proportional to the pressure of the gas. Using M1

$$- \frac{dp}{dt} = kp$$

This is the equation for a first-order reaction, k being the rate constant. Some examples of first-order heterogeneous reactions of this type are as follows:

$$2AsH_3 \xrightarrow{\text{glass}} 2As + 3H_2$$

$$2N_2O \xrightarrow{\text{gold}} 2N_2 + O_2$$

$$2HI \xrightarrow{\text{platinum}} H_2 + I_2$$

The last two reactions should be noted, because, when occurring homogeneously, they are often quoted as standard examples of second-order reactions.

Case 2.   *The surface is nearly covered*

In this case θ is almost unity and equation 8.4 becomes

$$k_1 = k_2 (1 - \theta) p$$

Rearranging

$$(1 - \theta) = \frac{k_1}{k_2 p} \tag{8.6}$$

i.e. the fraction of the surface remaining uncovered is inversely proportional to the pressure of the gas.

This may be illustrated with reference to the catalytic hydrogenation of ethene on a copper catalyst at $0\,^\circ$C. It is found experimentally that the rate is directly proportional to the concentration of hydrogen, but inversely proportional to the concentration of ethene, i.e.

$$\text{rate} \propto \frac{[H_2]}{[C_2H_4]} \qquad (8.7)$$

From the definition of order of reaction (p. 8) equation 8.7 is a zero-order rate law. It may be interpreted by assuming that the hydrogen is slightly adsorbed whereas the ethene is strongly adsorbed, and covers most of the surface. Since the surface is almost covered by ethene, the rate-determining factor will be the adsorption of hydrogen. This will depend on the pressure of the hydrogen (from equation 8.5) and on the fraction of surface *not* covered by ethene. As seen in equation 8.6, the latter will be inversely proportional to the pressure of ethene and so the rate of reaction is proportional to $[H_2]$ and to $1/[C_2H_4]$, in agreement with the experimental result 8.7.

The decomposition of ammonia on platinum is also of this type, the rate being proportional to the pressure of ammonia and inversely proportional to the pressure of hydrogen, which is therefore deduced to be strongly adsorbed.

Case 3. *The surface is saturated*

When saturation occurs the amount of adsorbed gas is constant and cannot be increased by increasing the pressure. The rate of reaction is, therefore, constant and a zero-order reaction is observed.

Two examples of this type of reaction are the high-temperature catalytic decomposition of ammonia on tungsten or molybdenum, or the catalytic decomposition of hydrogen iodide on gold:

$$2NH_3 \xrightarrow{\text{tungsten}} N_2 + 3H_2$$

$$2HI \xrightarrow{\text{gold}} H_2 + I_2$$

In zero-order reactions of this type, if the pressure is reduced sufficiently a stage will eventually be reached at which the surface is only slightly covered with adsorbed gas and equation 8.5 will apply. The order of the reaction will therefore change from zero to one on decreasing the pressure. At intermediate pressures, when the surface is fractionally covered, the reaction will have a fractional order.

Case 4.   *Adsorption with dissociation*

In some hydrogenation reactions, hydrogen molecules are dissociated into atoms (see Section 8.2), each of which occupies one site on the surface. Calling the free surface sites S, and the sites with adsorbed hydrogen S–H, then the following equilibrium is set up

$$H_2 + 2S \rightleftharpoons 2S-H \tag{8.8}$$

The equilibrium constant (K) for equation 8.8 is

$$K = \frac{[S-H]^2}{[H_2][S]^2} \tag{8.9}$$

Now the fraction of surface covered (θ) is proportional to $[S-H]$ whilst the fraction of surface uncovered $(1 - \theta)$ is proportional to $[S]$. Hence equation 8.9 can be written

$$K = \frac{\theta^2}{[H_2](1 - \theta)^2}$$

or

$$\theta = (1 - \theta)K^{1/2}[H_2]^{1/2} \tag{8.10}$$

When the surface coverage is small, θ is proportional to the square root of the pressure of hydrogen, and the reaction has an order of one-half with respect to hydrogen. As in Case 3, if the pressure is increased the order decreases to zero as the surface becomes saturated. This type of kinetic behaviour has been observed in hydrogenations on nickel catalysts.

The above examples show that, in suitable cases, simple kinetic behaviour may be found in heterogeneous reactions. In general, the situation is more complex since the adsorption may be intermediate between Case 1 and Case 2. Furthermore the products of the reaction may be adsorbed. For example, in the catalytic oxidation of sulphur dioxide on a platinum catalyst (the Contact Process)

$$2SO_2 + O_2 \xrightarrow{\text{platinum}} 2SO_3$$

the following rate laws are obeyed.

When sulphur dioxide is in excess

$$\frac{d[SO_3]}{dt} = \frac{k[O_2]}{[SO_3]^{1/2}}$$

When oxygen is in excess

$$\frac{d[SO_3]}{dt} = \frac{k[SO_2]}{[SO_3]^{1/2}}$$

The occurrence of the concentration of sulphur trioxide in the denominator in the rate laws shows that it is being strongly adsorbed and is inhibiting the reaction.

8.4   Mechanism of Heterogeneous Catalysis

The function of a catalyst is to provide a mechanism of reaction with a lower energy of activation than is found in the corresponding homogeneous reaction, thus making the reaction on the surface faster than that in the gas phase.

A catalyst lowers the energy of activation by adsorbing the gas on certain positions on the surface. This idea enables a simplified explanation to be given for the specificity of catalytic action mentioned on p. 96. It is probable that two or more neighbouring sites in the surface hold the molecule by a process of co-operative adsorption. As a result, there is stretching of some of the bonds of the adsorbed molecule which is thus taken part of the way to the transition state. In other words, the activation energy is lowered.

Different catalysts can adsorb a given molecule in different ways, and so produce different products in the subsequent reaction. The manner and extent of the adsorption is critically dependent on the spacing of the surface sites, and thus on the chemical nature of the surface.

The data in *Table 8.1*, in which the energies of activation of corresponding homogeneous and heterogeneous reactions are given, illustrate the lowering of activation energy by a catalyst.

It must be stressed that the above treatment is based on the assumption that the solid surface is uniform. This is very seldom the case.

Table 8.1

Reaction	Catalyst	Energy of activation/ kJ mol^{-1}	
		Heterogeneous	Homogeneous
$2HI \rightarrow H_2 + I_2$	Platinum	59	184
$2N_2O \rightarrow 2N_2 + O_2$	Gold	121	245
$2NH_3 \rightarrow N_2 + 3H_2$	Tungsten	163	335
$CH_4 \rightarrow C + 2H_2$	Platinum	230	335

8.5 Active Site Theory

From his study of catalysed reactions, H.S. Taylor developed the *active site theory* in which it is postulated that the solid surface is not uniform, and that a catalysed chemical reaction can occur only on a few specially favoured sites. On the other hand, adsorption of gas molecules can occur over the whole surface, so that the majority of adsorbed molecules may be held on non-active sites. The rate of reaction will be proportional to the amount of gas adsorbed on to the active sites.

Strong evidence for this theory comes from an investigation of the *poisoning* of catalysts, in which a trace of added substance removes the catalytic activity. Thus, in the hydrogenation of ethene to ethane, on a copper catalyst, a trace of mercury (far less than that required to form a layer one atom thick on the surface) is sufficient to inhibit the reaction. This is because the mercury is preferentially adsorbed on the active sites.

Even more striking is the fact that a catalyst may be 'poisoned' with respect to one reaction and yet capable of catalysing another. For example, colloidal platinum will catalyse the hydrogenation of ketones and of aromatic nitro compounds. Small amounts of carbon disulphide will prevent the hydrogenation of ketones, but the partially poisoned platinum is still capable of catalysing the hydrogenation of nitrobenzene. Further small additions of carbon disulphide completely remove the catalytic activity.

Additional evidence for the non-uniformity of catalyst surfaces comes from the effect of sintering. Many catalysts can be completely deactivated by heating to a temperature below their melting point. Heating will tend to make the surface atoms more mobile and will thus tend to even out local distortions of the surface, including those distortions which constitute the active sites. Part of the reduction in catalytic power on heating can be accounted for by the fact that the surface area of the catalyst tends to decrease on sintering, so that less gas is adsorbed. Nevertheless, the loss in catalytic activity can be many times greater than would be produced by the reduction in surface area.

Kinetic studies also lead to the conclusion that catalytic surfaces are non-uniform. An investigation of the reaction

$$CO_2 + H_2 \rightarrow CO + H_2O$$

on a platinum surface shows that when an excess of carbon dioxide is used it slows down the reaction. The kinetics of the reaction can be described by the equation

$$\text{rate} \propto \frac{\text{partial pressure of hydrogen}}{\text{partial pressure of carbon dioxide}} \qquad (8.11)$$

An examination of equation 8.6 shows that the above result implies that the carbon dioxide is strongly adsorbed and covers most of the surface, and that the rate is determined by the adsorption of hydrogen on that small part of the surface not covered by carbon dioxide. However, direct measurements of the amounts of carbon dioxide and hydrogen adsorbed on the platinum catalyst show that far more hydrogen is adsorbed than carbon dioxide.

These results can be easily explained by the active site theory if equation 8.11 is interpreted as meaning that the carbon dioxide is preferentially adsorbed on the active sites, which are only a small fraction of the total surface. The amount of hydrogen adsorbed on the non-active sites can thus be much greater than the amount of carbon dioxide adsorbed on the active sites.

It is interesting to compare the rates of decomposition of formic acid on various surfaces, shown in *Table 8.2*. Thus on glass and

Table 8.2 $HCOOH \rightarrow H_2 + CO_2$

Surface	Energy of activation/ kJ mol^{-1}	Relative rate
Glass	102	1
Gold	98	40
Silver	130	40
Platinum	92	2000
Rhodium	104	10 000

on rhodium the rates differ by a factor of 10 000 while the energies of activation are almost equal. This can reasonably be interpreted by assuming that rhodium has far more active sites per unit area than has glass.

A quantitative theory of heterogeneous catalysis has not yet been developed. This is not surprising in view of the complex nature of the solid surface. However, the application of the simple idea of a monomolecular layer of chemisorbed gas on the surface, combined with the Langmuir isotherm, equation 8.4, does enable an account to be given of the kinetics of heterogeneous reactions in a few cases.

9

HOMOGENEOUS CATALYSIS

Homogeneous catalysis differs in two ways from heterogeneous catalysis (discussed in Chapter 8). First, there are no surface effects so that the rate of reaction is independent of the chemical nature of the walls of the container and of the ratio of surface area to volume. Secondly, the catalyst undergoes chemical reaction with one of the reactants and subsequently is regenerated to its original chemical form. Also the kinetic effect of the catalyst can be measured directly as its concentration can be varied in a known way.

The feature that is in common with heterogeneous reactions is that a mechanism of reaction with a reduced activation energy is provided. This is true in all catalysed reactions.

The reaction and regeneration of the catalyst can be seen clearly in some gas-phase examples, which will be considered first. Following this is a discussion of homogeneous catalysis in solution and finally there are sections dealing with various aspects of acid–base catalysis.

9.1 Homogeneous Catalysis in the Gas Phase

A well-known example of this type of catalysis occurs in the Lead Chamber Process for the manufacture of sulphuric acid. Sulphur dioxide is not directly oxidized to sulphur trioxide, but this oxidation does occur in the presence of nitrogen dioxide:

$$SO_2 + NO_2 \rightarrow SO_3 + NO$$
$$NO + \tfrac{1}{2}O_2 \rightarrow NO_2$$

Overall reaction $\quad SO_2 + \tfrac{1}{2}O_2 \rightarrow SO_3$

Nitrogen dioxide also catalyses the oxidation of carbon monoxide to carbon dioxide:

$$CO + NO_2 \rightarrow CO_2 + NO$$
$$NO + \tfrac{1}{2}O_2 \rightarrow NO_2$$

Overall reaction $\quad CO + \tfrac{1}{2}O_2 \rightarrow CO_2$

These reactions clearly show the role of the catalyst, which participates in the first stage and is then regenerated in the second stage of the reaction. Iodine vapour accelerates by a factor of several hundred the rate of thermal decomposition of aldehydes and ethers. In the case of ethanal the mechanism is

$$CH_3CHO + I_2 \rightarrow CH_3I + HI + CO$$
$$CH_3I + HI \rightarrow CH_4 + I_2$$

Overall reaction $CH_3CHO \rightarrow CH_4 + CO$

This mechanism can be compared with the uncatalysed thermal decomposition of ethanal described on p. 133.

The amount by which the activation energy is lowered by the catalyst can be seen by considering the thermal decomposition of di-isopropyl ether. For the uncatalysed reaction, the energy of activation is 255 kJ mol^{-1}. In the presence of an iodine catalyst, this value drops to 119 kJ mol^{-1}.

An interesting example of gas-phase catalysis with unusual kinetics is the decomposition of ozone to oxygen in the presence of dinitrogen pentoxide. Using the idea of a rapid pre-equilibrium as postulated by Ogg (Section 7.6) the catalysis can be explained as follows. The first stage is

$$N_2O_5 \rightleftharpoons NO_2 + NO_3$$

which is followed by oxidation of the nitrogen dioxide by the ozone

$$NO_2 + O_3 \rightarrow NO_3 + O_2$$
$$NO_3 \rightarrow NO_2 + \tfrac{1}{2}O_2$$

Overall reaction $\quad O_3 \rightarrow \tfrac{3}{2}O_2$

If the stationary state hypothesis is applied to the NO_2 and

NO_3 molecules, then the rate law required by the above mechanism is

$$\text{rate} = k[O_3]^{2/3}[N_2O_5]^{2/3} \tag{9.1}$$

Verification of equation 9.1 will provide the reader with useful practice in the application of the stationary state hypothesis. This rather unusual four-thirds-order rate law has in fact been observed in studying the kinetics of this decomposition at various initial concentrations of dinitrogen pentoxide. This is another example of using kinetic studies for elucidating reaction mechanisms.

9.2 Homogeneous Catalysis in Solution

In principle, catalysis in solution is similar to that in the gas phase. The catalyst reacts with one of the reactants, and then by regenerative recycling is restored to its original chemically active state. To illustrate these ideas consider the catalysis of the very slow reaction between bromate and arsenic(III) ions in aqueous solution. Thermodynamically, the reaction is very feasible in acidic solution, but the rate of the reaction

$$BrO_3^- + 3As(III) + 6H^+ \rightarrow Br^- + 3As(V) + 3H_2O$$

at room temperature is exceedingly slow.

The presence of a trace of osmium(VIII), sometimes referred to as osmic acid, produces a dramatic increase in rate, because a reaction route with a much lower activation energy has been made available. Kinetic studies show that in dilute aqueous acidic solution the rate can be expressed by

$$\text{rate} = k[Os(VIII)][As(III)] \tag{9.2}$$

This second-order rate law indicates a rate-determining stage between Os(VIII) and As(III) which does not involve the BrO_3^- and the H^+ ions. Their contribution comes in the regenerative stage which is not rate-determining. Thus the mechanism of the reaction can be interpreted by the equations

$$Os(VIII) + As(III) \xrightarrow[\text{controlling}]{\text{rate}} Os(VI) + As(V)$$

$$BrO_3^- + 2H^+ + Os(VI) \rightarrow Os(VIII) + BrO_2^- + H_2O$$

$$BrO_2^- + 2H^+ + As(III) \rightarrow BrO^- + H_2O + As(V)$$

$$BrO^- + 2H^+ + As(III) \rightarrow Br^- + H_2O + As(V)$$

Overall reaction $BrO_3^- + 6H^+ + 3As(III) \rightarrow Br^- + 3H_2O + 3As(V)$

The first two stages constitute the regenerative catalytic cycle. The exact nature of the reacting species cannot be elucidated purely from kinetic studies. The reader is referred to Section 7.7, in which the mechanism of the bromide–bromate reaction is discussed.

Each stage in the mechanism proposed above can be regarded as bimolecular with hydrogen ions associated with the negative ions by ion-pairing or by formal bonding during rapidly established equilibrium processes.

An interesting comparison can be made between the catalytic cycle of this reaction and that of the osmium catalysed reaction between cerium(IV) and arsenic(III) ions. In acidic solution and in the presence of excess Ce(IV) the rate law is

$$rate = k[Os(VIII)][As(III)] \tag{9.3}$$

Hence by similar reasoning to that used above a cycle of catalytic regeneration can be proposed, namely

$$Os(VIII) + As(III) \xrightarrow[\text{controlling}]{\text{rate}} Os(VI) + As(V)$$

$$2Ce(IV) + Os(VI) \rightarrow Os(VIII) + 2Ce(III)$$

Overall reaction $2Ce(IV) + As(III) \rightarrow 2Ce(III) + As(V)$

9.3 Acid–Base Catalysis: Brönsted Theory

The expression *acid–base catalysis* refers to a homogeneous reaction in solution in which either an acid or a base acts as a catalyst. An example of such a reaction is the hydrolysis of esters

$$R^1CO_2R^2 + H_2O \rightarrow R^1CO_2H + R^2OH$$

which is catalysed both by acids and by bases in aqueous solution.

Many of the quantitative studies of acid–base catalysis were carried out by Brönsted. He discovered that the observed kinetic effects could be explained if it was recognized that

(a) an acid is a substance capable of donating a proton to another molecule;

(b) a base is a substance that can accept a proton from an acid.

These are referred to as the *Brönsted definitions of the terms 'acid' and 'base'* and are used in this chapter. The Brönsted scheme can be represented as

$$A \rightleftharpoons B + H^+ \qquad (9.4)$$

where the acid A and the base B are called a *conjugate acid-base pair*, since they differ only by a proton.

It may not at first sight be obvious that this idea differs significantly from the older view that acidity is proportional to concentration of hydrogen ions. Indeed, in dilute aqueous solution there is little difference between the two definitions. The difference can be seen, however, by considering absolutely dry ethanoic acid (100.00 per cent glacial acetic acid). This is a non-conductor of electricity and contains no hydrogen ions, yet it is a powerfully acidic solvent because the undissociated acetic acid molecules can act as an acid by protonating a base. Similarly, pure liquid ammonia is a strongly basic solvent, although it contains no hydroxide ions. The ammonia molecule can accept protons from acids, forming ammonium ions. In aqueous solutions the acetate ion is a base and the ammonium ion is an acid.

Water plays a key role in acid–base catalysis. Because of its polar nature and high dielectric constant it is a good solvent for electrolytes. In addition, water takes part in acid–base reactions both as an acid and as a base. A solvent which can act as both an acid (proton donor) and a base (proton acceptor) is called an *amphoteric solvent*. Such a solvent may be represented as the molecule XH. An amphoteric solvent will take part in the following equilibria with an acid (HA) or with the conjugate base of the acid (A^-)

$$HA + XH \rightleftharpoons XH_2^+ + A^-$$
$$\text{acid} \quad \text{base} \quad \text{acid} \quad \text{base}$$

$$A^- + XH \rightleftharpoons X^- + HA$$
$$\text{base} \quad \text{acid} \quad \text{base} \quad \text{acid}$$

Water, methanol and ethanol are examples of amphoteric solvents.

In aqueous solution the protonated solvent molecule is the H_3O^+ ion. For convenience, this has been abbreviated to H^+ in this chapter. It must be emphasized that the unsolvated proton cannot exist in solution. The existence of H_3O^+ has been shown clearly by nuclear magnetic resonance studies of the solidified hydrates formed on adding one mole of water to one mole of sulphuric, nitric and perchloric acids, respectively. In each case the spectrum shows that there are three protons in an equilateral triangle, corresponding to the three protons of H_3O^+, the structure of which is thought to be a very flat pyramid. For example, $H_2SO_4 \cdot H_2O$ is actually $H_3O^+HSO_4^-$. It is noteworthy that the monohydrate of perchloric acid $(H_3O^+ClO_4^-)$ is isomorphous with ammonium perchlorate, which undoubtedly has the structure $NH_4^+ClO_4^-$.

An essential idea in acid–base catalysis is that *both* an acid and a base must be present. It is not sufficient to protonate the substrate molecule to carry out effective acid catalysis. In addition, the proton must be removed from the reaction intermediate at some stage by a base (often the solvent).

The validity of this idea has been demonstrated by experiments on the mutarotation of 2,3,4,6-tetramethyl glucose which is appreciably soluble in both dry pyridine and dry cresol. Pyridine is a strongly basic solvent whereas cresol is a strongly acidic solvent. The rate of mutarotation in either pure solvent is extremely slow. However, in a mixture of the two solvents the reaction is very rapid, showing that both an acid and a base are necessary for the catalysis.

9.4 Acid–Base Catalytic Constants

Acid–base catalytic constants measure the effectiveness of the acid–base catalysts. If a number of acids and bases are present in the solution, each species may contribute to the observed rate constant k. If a solution contains a weak acid HA and its conjugate base A^- then the general expression for the rate constant is

$$k = k_0 + k_{H^+}[H^+] + k_{OH^-}[OH^-] + k_{HA}[HA] + k_{A^-}[A^-]$$

$$(9.5)$$

where k_0 is the rate constant of the uncatalysed reaction, and k_{H^+}, k_{OH^-}, k_{HA} and k_{A^-} are the catalytic constants of H^+, OH^-, HA and A^-, respectively.

9.5 Specific Acid–Base Catalysis

Although in acid–base catalysis the presence of both acid and base are necessary, for some reactions the rate is proportional only to the concentration of H^+. This phenomenon is called *specific acid catalysis*, the implication of which is that the catalytic constant of the hydrogen ion is much larger than that of the other acids and bases in the reaction. Under these circumstances equation 9.5 becomes

$$k = k_{H^+}[H^+] \tag{9.6}$$

Taking logarithms of equation 9.6

$$\boxed{\log k = \log k_{H^+} - pH} \tag{9.7}$$

using the approximation $pH = -\log[H^+]$. $\tag{9.8}$

It will be seen from equation 9.7 that a plot of $\log k$ against pH is a straight line of slope -1.00. Similarly in specific base catalysis a reaction is catalysed only by OH^- ions, and so equation 9.5 becomes

$$k = k_{OH^-}[OH^-] \tag{9.9}$$

or

$$\log k = \log k_{OH^-} + \log[OH^-] \tag{9.10}$$

The ionic product principle gives $K_W = [OH^-][H^+]$ so that

$$\log[OH^-] = \log K_W - \log[H^+] \tag{9.11}$$

Putting 9.11 into 9.10 and using 9.8

$$\log k = \log k_{OH^-} + \log K_W + pH \tag{9.12}$$

The plot of $\log k$ against pH is, in this case, a straight line of slope 1.00.

The effect of pH on the rates of acid–base reactions can best be shown in a diagram, as in *Figure 9.1*. Curves (a) and (b) show reactions specifically catalysed by H^+ and OH^- ions, respectively. In reactions (such as the hydrolysis of esters) subject to both acid and base catalysis there is a minimum in the rate of

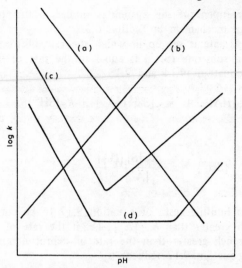

Figure 9.1 Effect of pH on reactions undergoing specific acid–base catalysis:
(a) inversion of sucrose (b) depolymerization of diacetone alcohol (c) hydrolysis
of esters (d) mutarotation of glucose

reaction in neutral solution when $[H^+] = [OH^-] = 10^{-7}$ mol l^{-1}
(at $25\,^{\circ}C$). Such behaviour is shown in curve (c) of *Figure 9.1*.
In curve (d) the rate of the uncatalysed reaction is significant,
and catalysis only becomes noticeable at high and low pH values.
The corresponding rate law is

$$k = k_0 + k_{H^+}[H^+] + k_{OH^-}[OH^-] \qquad (9.13)$$

9.6 General Acid–Base Catalysis

General acid or base catalysis involves a reversible reaction between
a substrate molecule S, which for instance could be glucose, and an
acid or base molecule. Consider for example a reaction catalysed
by the acid HA. The equation for the reaction is shown in
equation 9.14

$$S + HA \underset{k_{-1}}{\overset{k_1}{\rightleftharpoons}} SH^+ + A^- \qquad (9.14)$$

$$SH^+ + H_2O \overset{k_2}{\to} \text{reaction products} \qquad (9.15)$$

The kinetic treatment of this system is similar to that used in the Lindemann mechanism in Section 6.5.

A stationary state is set up in which the rate of formation of the protonated substrate (SH^+) is equal to the rate of its destruction. From equations 9.14 and 9.15

$$k_1 [S] [HA] = k_{-1} [SH^+] [A^-] + k_2 [SH^+] \qquad (9.16)$$

or

$$[SH^+] = \frac{k_1 [S] [HA]}{k_{-1} [A^-] + k_2} \qquad (9.17)$$

There are two limiting cases of equation 9.17 to be considered. If k_2 is much greater than $k_{-1} [A^-]$, i.e. if the rate of product formation is much greater than the rate of deprotonation of $[SH^+]$, then equation 9.17 reduces to

$$[SH^+] = k_1 [S] [HA] / k_2 \qquad (9.18)$$

Under these conditions the reaction is subject only to general acid catalysis by the Brönsted acid HA. The overall reaction rate is proportional to $[SH^+]$ and hence to the concentrations of the substrate and of the molecular form of the acid. If k_2 is much less than $k_{-1} [A^-]$ then equation 9.17 reduces to

$$[SH^+] = \frac{k_1 [S] [HA]}{k_{-1} [A^-]} \qquad (9.19)$$

However, the acid HA undergoes the reversible ionization

$$HA \rightleftharpoons H^+ + A^- \qquad (9.20)$$

for which the equilibrium constant (K) may be written

$$K = [H^+] [A^-] / [HA] \qquad (9.21)$$

combining equation 9.21 with 9.19

$$[SH^+] = k_1 [S] [H^+] / k_{-1} K \qquad (9.22)$$

In this case the reaction will be subject to specific acid catalysis only, as the reaction rate is strictly proportional to $[H^+]$ even in the presence of variable amounts of HA and A^-.

The existence of both general and specific acid–base catalysis can be explained in terms of kinetic arguments of the type just used. It was the existence of the phenomenon of general catalysis that led to the previously stated Brönsted definitions of acid and base.

In acid catalysis, the first stage of the reaction is the protonation of the substrate molecule. It might therefore be expected that the effectiveness of an acid catalyst should be related to the protonating power and hence the strength of the acid. Brönsted showed that this was indeed the case. There is a correlation between the catalytic constant of an acid, in a particular reaction, and its ionization constant. The *Brönsted relation* states that the logarithm of the catalytic constant plotted against the logarithm of the ionization constant gives a straight line.

A similar argument applies to base catalysis, where the ionization constant of the base (K_B) is inversely proportional to the ionization constant (K_A) of the conjugate acid. This follows from the fact that in aqueous solution

$$K_A \cdot K_B = K_W \qquad (9.23)$$

where K_W is the ionic product of water.

Typical kinetic evidence from which the existence of general base catalysis is deduced can be seen by considering the decomposition of nitramide. A set of experiments was carried out in an acetic acid/sodium acetate buffer solution. This was weakly acid and so the concentration of hydroxide ions was negligible. The observed rate of reaction was not related to the concentration of hydrogen ions or of acetic acid molecules, but it varied linearly with the concentration of acetate ions. This reaction is therefore catalysed by acetate ions, which act as a Brönsted base. The catalytic constant of the acetate ion is determined by this experiment. Amines also catalyse this reaction.

These ideas are illustrated in *Table 9.1* in which details of the base-catalysed decomposition of nitramide are given. The kinetic studies were carried out in various buffer solutions of the type Na^+A^-/HA, and the catalytic constant of the base was determined by the method described above. The table shows clearly that the hydroxide ion is catalytically much more effective than the other basic ions. It is only in acidic solutions where the concentration of hydroxide ions is reduced to a very low level that general base catalysis can be observed.

General acid–base catalysis was first established from a careful study of the kinetics of the iodination of acetone in aqueous

Table 9.1 Base-catalysed decomposition of nitramide,
$NH_2NO_2 \rightarrow N_2O + H_2O$

Base catalyst	Conjugate acid	Ionization constant of acid (25 °C)/ mol l^{-1}	Catalytic constant at 15 °C/ l mol^{-1} min^{-1}
Hydroxide	Water	1.0×10^{-14}	1.0×10^6
Acetate	Acetic	1.8×10^{-5}	0.504
Chloroacetate	Chloroacetic	1.4×10^{-3}	1.6×10^{-2}
Dichloroacetate	Dichloroacetic	5.0×10^{-2}	7.0×10^{-4}

acetate buffer solutions. It was found experimentally that in a
given buffer solution the rate of reaction is proportional to the
concentration of acetone, and is accurately given by the equation

$$\text{rate} = k[\text{acetone}] \qquad (9.24)$$

where k is the experimentally determined first-order rate constant
for the reaction. It should be remembered (see p. 76) that this
reaction is of zero order with respect to iodine.

The value of the rate constant k in equation 9.24 varies with
the composition of the buffer solution, in accordance with the
general acid–base equation shown in equation 9.5. The following results have been found at 25 °C:

$$k_0 = 4.6 \times 10^{-10} \text{ s}^{-1}$$
$$k_{H^+} = 2.7 \times 10^{-5} \text{ l mol}^{-1} \text{ s}^{-1}$$
$$k_{OH^-} = 0.25 \text{ l mol}^{-1} \text{ s}^{-1}$$
$$k_{HA} = 8.3 \times 10^{-8} \text{ l mol}^{-1} \text{ s}^{-1}$$
$$k_{A^-} = 2.5 \times 10^{-7} \text{ l mol}^{-1} \text{ s}^{-1}$$

The catalytic constant of H^+ is much larger than that of HA and
the catalytic constant of OH^- is much larger than that of A^-.
Consequently, in aqueous solution the catalytic effects of H^+ and
OH^- frequently overwhelm those due to other acids or bases in
the solution. General catalysis can only be demonstrated by
experiments carefully conducted in buffer solutions in which the
concentrations of H^+ and OH^- are accurately controlled.

10

ENZYME CATALYSIS

10.1 Enzymes

Enzymes differ from other catalysts in that they are macromolecules synthesized by living organisms. Chemically all enzymes are proteins with molecular weights ranging from 10^4 to 10^6. Their activity sometimes depends on the presence of non-protein substances (e.g. metallic ions) known as co-enzymes. Their most remarkable features are their great catalytic activity and their specificity. For example, the enzyme urease will catalyse the hydrolysis of urea

$$CO(NH_2)_2 + H_2O \rightarrow CO_2 + 2NH_3$$

and no other reaction.

Enzymes are classified according to the reaction or type of reaction which they catalyse. Thus urease is an enzyme in a group called the hydrolases. Alcohol dehydrogenase, as its name suggests, facilitates the conversion of alcohols to aldehydes and is a member of the oxidoreductases. Decarboxylase is one of the lyases which catalyses reactions of the type

$$R^1CHR^2CO_2H \rightarrow R^1CH_2R^2 + CO_2$$

In comparison with other catalysts, enzymes are more effective by many orders of magnitude.

In hydrolysis of urea catalysed by hydrogen ions the activation energy is 104 kJ mol^{-1}. In the same reaction catalysed by urease the activation energy is only 29 kJ mol^{-1}. At equal molarities urease is about 10^{12} times more effective than acid catalysis. A similar comparison shows the enzyme catalase to be about 10^5 times more effective than ferrous ion in the decomposition of hydrogen peroxide.

The high activity typical of enzyme catalysis enables reactions to proceed rapidly in the presence of a very low concentration of enzyme and under mild conditions, e.g. in the human body. Typical enzyme concentrations for most systems range from 10^{-8} to 10^{-10} mol l^{-1}, while the concentration of the substrate with which the enzyme reacts is usually greater than 10^{-6} mol l^{-1}.

In passing it is interesting to note that the size of the enzyme molecule relative to substrate molecule is very large. The substrate may represent a portion of about 1% of the total enzyme-substrate mass, in spite of the higher molar concentration of substrate.

Despite their biological origins and the special role they have within living systems, in many respects enzymes have characteristics similar to catalysts from non-living sources. Their function is not restricted to the confines of living cells. They can be extracted from source, purified and crystallized and then used advantageously as particularly selective and active catalysts under laboratory or industrial conditions. However, enzymes, like all proteins, undergo a process called denaturation when subjected to elevated temperatures or extremes of pH.

10.2 Enzyme Structure

Denaturation results in the disruption of the highly ordered three-dimensional structure of the enzyme and inevitable loss of catalytic activity.

The unique function of an enzyme resides in the active site that is established by a particular spatial configuration of the protein peptide chain. The complex folding of the peptide chain forms a cleft in which the substrate binding points are critically placed. Enzymes distinguish between various possible substrates by requiring a substrate to fit the shape of the active site and for it to attach at these specific binding points.

The enzyme structure is not rigid and, by limited flexing and distortion of the occupied cleft, the bond-breaking and bond-making processes are assisted. Essentially this is how the enzyme lowers the activation energy of the reaction it is catalysing.

Figure 10.1 is a schematic diagram of the structure of the enzyme chymotrypsin. The folding line represents the folding of the peptide chain that establishes the region of the active site of the enzyme which is occupied by a substrate ester. The peptide chain consists of links made up of amino-acid residues. These links are numbered according to their sequence in the chain. The relative spatial positions of four important amino-acid residues are shown together with their number in the chain sequence. Thus the essential structure of the chain is

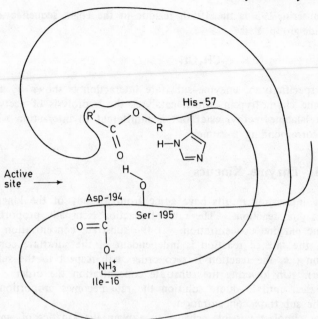

Figure 10.1   Schematic diagram of the active site of the enzyme chymo-trypsin.   An ester substrate is schematically shown located in the active site and undergoing hydrolysis assisted by the attachment points histidine-57 (His-57) and serine-195 (Ser-195).   Aspartate-194 (Asp-194) and isoleucine-16 (Ile-16) contribute to the conformation of the active site

$$\cdots NH\!-\!CH\!-\!CO\!-\!NH\!-\!CH\!-\!CO\cdots$$
$$\mid \qquad\qquad \mid$$
$$Y \qquad\qquad Y$$

where

$$-NH\!-\!CH\!-\!CO-$$
$$\mid$$
$$Y$$

is referred to as an amino-acid residue, $-CO-NH-$ is a peptide linkage and Y is a side-group characteristic of each amino acid. For example, His-57, i.e. histidine-57, is the 57th residue in the peptide chain in which the side-group Y is while serine-195 is the 195th residue in the chain sequence and its side-group Y is

$$-CH_2OH$$

The specificity of enzyme–substrate interaction is shown by the enzyme chymotrypsin, which catalyses the hydrolysis of acetyl-L-phenylalanine methyl ester but is indifferent to interaction with the corresponding D-isomer.

10.3 Enzyme Kinetics

Some interesting results have come from a study of the kinetics of enzyme reactions. The rate of reaction is usually proportional to the enzyme concentration. If the substrate concentration is high, the rate of reaction is independent of the substrate concentration (i.e. the reaction is zero order with respect to the substrate). On lowering the substrate concentration the order increases, until in dilute solution the rate becomes proportional to the substrate concentration.

The simplest possible scheme to explain the kinetics of single substrate reactions is the *Michaelis–Menten mechanism*. The enzyme is denoted by E and the substrate by S:

$$E + S \underset{k_{-1}}{\overset{k_1}{\rightleftharpoons}} ES \tag{10.1}$$

$$ES \overset{k_2}{\rightarrow} E + \text{products} \tag{10.2}$$

This is similar in mathematical form to the mechanism used in equations 9.14 and 9.15 to explain the existence of specific and general acid catalysis.

Assuming a steady state to be set up, the rate of formation of ES is taken to equal the rate of its removal, i.e.

$$k_1 [E] [S] = k_{-1} [ES] + k_2 [ES] \tag{10.3}$$

Since the enzyme exists either in the free form (E) or the combined form (ES), the total enzyme $[E]_0$ originally added is

$$[E]_0 = [E] + [ES] \tag{10.4}$$

Replacing [E] in equation 10.3 by $[E]_0 - [ES]$ (from equation 10.4)

$$k_1\{[E]_0 - [ES]\}[S] = (k_{-1} + k_2)[ES]$$

Rearranging

$$[ES] = \frac{k_1[E]_0[S]}{k_{-1} + k_2 + k_1[S]}$$

From equation 10.2 the rate of reaction is $k_2[ES]$. Hence

$$\text{rate} = k_2[ES] = \frac{k_1k_2[E]_0[S]}{k_{-1} + k_2 + k_1[S]} \tag{10.5}$$

Dividing the numerator and denominator of the right-hand side of equation 10.5 by k_1 gives

$$\text{rate} = \frac{k_2[E]_0[S]}{K_m + [S]} \tag{10.6}$$

where

$$K_m = (k_{-1} + k_2)/k_1 \tag{10.7}$$

The constant K_m is known as the *Michaelis constant*. In general it is not an equilibrium constant but a 'steady-state' constant. Only with reactions for which k_2 is much less than k_{-1} is K_m numerically equal to the dissociation constant of the equilibrium

$$ES \rightleftharpoons E + S$$

Equation 10.6 shows that the rate of reaction is directly proportional to enzyme concentration; it also gives the dependency of rate on substrate. Provided the total enzyme concentration is unchanged, a plot of the initial reaction rate against the initial substrate concentration has the characteristic shape illustrated in *Figure 10.2*. At very low substrate concentrations, [S] is much less than K_m and the plot obeys

$$\text{rate} = k_2[E]_0[S]/K_m \tag{10.8}$$

which shows first-order dependency in S. Under more typical conditions, when [S] is much greater than K_m

$$\text{rate} = k_2[E]_0 \tag{10.9}$$

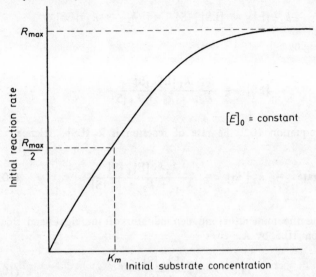

Figure 10.2 Initial rate of an enzyme-catalysed reaction obeying equation 10.6

When this condition applies, a maximum or limiting rate of reaction (R_{max}) is attained, with effectively all of the enzyme in the form of the enzyme–substrate complex. Thus

$$R_{max} = k_2 [E]_0 \qquad (10.10)$$

and the reaction shows zero-order kinetics with respect to the substrate S.

Substituting equation 10.10 into equation 10.6 gives

$$\text{rate} = \frac{R_{max} [S]}{K_m + [S]} \qquad (10.11)$$

This situation is similar to the case in heterogeneous catalysis where all the active sites of a catalyst surface are saturated (see Section 8.3, Case 3).

The Michaelis constant has the dimension of concentration and is equal to that concentration of substrate which gives an initial rate of $R_{max}/2$. The ratio of $R_{max}/[E]_0$ is called the *turnover number*. It has the dimension of reciprocal time and indicates the number of substrate molecules converted to products per active site in unit time.

Although the Michaelis–Menten mechanism, discussed above, has considered the simplest of reaction schemes, it interprets with remarkable consistency the kinetics generally observed in more complex enzyme catalysis. (The reader is referred to the Lindemann mechanism of unimolecular gas reactions, Section 6.5, that makes an interesting comparison with the Michaelis–Menten enzyme scheme.)

An illustrative example of enzyme kinetics is given in the description of the determination of the Michaelis constant (and related parameters) as a laboratory experiment.[*] The enzyme used in the experiment is chymotrypsin, which catalytically cleaves an amide bond in the substrate[†], liberating *p*-nitroaniline which absorbs strongly at 410 nm. At this wavelength, the enzyme, substrate and other products do not absorb appreciably, so the reaction can be conveniently followed using a recording spectrophotometer.

A set of experiments is carried out using the same initial enzyme concentration, but varying the initial concentration of substrate. In each experiment, the initial rate of reaction is found from the linear change of absorbance at 410 nm with time. The concentration is calculated from the absorbance using the known value (8200 $1 \, mol^{-1} \, cm^{-1}$) of the molar absorptivity of *p*-nitroaniline at this wavelength. The kinetic parameters K_m and k_2 (defined by equations 10.10 and 10.11) can be obtained from the rate data by graphical methods.

The Michaelis–Menten plot shows rate versus substrate concentration and gives a curved line. The Lineweaver–Burk plot shows the reciprocal of rate versus the reciprocal of the substrate concentration. The Hofstee plot shows rate versus rate/substrate concentration. The Woolf plot shows substrate concentration/rate versus substrate concentration. As required by equation 10.6, the latter three plots all give straight lines.

The value of the rate constant k_2 is easily obtained from equation 10.10 since the value of R_{max} has been found experimentally and the initial enzyme concentration is known. The reader is recommended to test the Woolf plot which is based upon rearrangement of equation 10.11 in the form

$$\frac{[S]}{rate} = \frac{[S]}{R_{max}} + \frac{K_m}{R_{max}}$$

The data in *Table 10.1* can be used for this purpose.

[*]HURLBUT, J.A. *et al.*, *J. Chem. Educ.*, **50**, 149 (1973)
[†]The substrate GPNA is *N*-glutamyl-L-phenylalanine *p*-nitroanilide

Table 10.1 Chymotrypsin catalysed cleavage of GPNA*

$[S]/10^{-4}$ mol l^{-1}	*Rate of change of absorbance*/min^{-1}	Rate/10^{-6} mol l^{-1} min^{-1}
2.5	0.018	2.2
5.0	0.031	3.8
10.0	0.048	5.9
15.0	0.058	7.1

*Initial enzyme concentration = 4×10^{-6} mol l^{-1}; cell width = 1 cm

The Woolf plot of $[S]$/rate versus $[S]$ gives a straight line of slope $1/R_{max}$ and intercept (at $[S] = 0$) of K_m/R_{max}. This is how the Michaelis constant (K_m) can be obtained.

The quoted values for the kinetic parameters derived from the data in *Table 10.1* are:

$$K_m = 1.1 \times 10^{-3} \text{ mol } l^{-1}$$

$$R_{max} = 1.3 \times 10^{-5} \text{ mol } l^{-1} \text{ min}^{-1}$$

$$k_2 = 3.2 \text{ min}^{-1}$$

10.4 Composition of Active Site: Chymotrypsin

Chymotrypsin is an enzyme that catalyses the hydrolysis of peptides, amides and esters. It is readily obtainable from beef pancreas, and easily purified by crystallization. The molecular weight is about 25 000 and the sequence of its 246 amino-acid residues has been elucidated by crystallography and other structural studies. The critical disposition of the binding points within its active site has been indicated by studying the effect on reaction rate of substrates with different structural features. With some substrates a stable enzyme intermediate can be isolated and a particular binding point identified.

Chymotrypsin treated with *p*-nitrophenylacetate produces a stable acetyl–enzyme intermediate. This intermediate can be isolated and the amino-acid sequence analysed. The analysis locates the acetyl label at serine-195, the 195th amino-acid residue along the peptide chain (see *Figure 10.1*).

The results of various kinetic studies, in particular those of experiments using pH variation and kinetic isotope effects, have indicated a histidine residue as another binding point. Methylation of histidine-57 leads to an inactive enzyme, which is good evidence for this residue as a substrate binding point.

Crystallographic structure determinations show histidine-57 and serine-195 in close spatial proximity, although widely separated in the chain sequence. The structure determination also shows how alkaline pH affects the enzyme specificity. Normally, the carboxylate ion of the aspartate-194 residue is held away from the active site by ion pairing with the NH_3^+ of the terminal residue isoleucine-16. Neutralization of $-NH_3^+$ allows the carboxylate group to swing into the cleft of the active site and affect specific substrate binding.

A reaction sequence for chymotrypsin-catalysed hydrolysis of specific substrate esters could be

$$E + R^1O-\overset{\overset{O}{\|}}{C}R^2 \underset{K_S}{\rightleftharpoons} ES \xrightarrow{k_2} R^2\overset{\overset{O}{\|}}{C}OE \xrightarrow[H_2O]{k_3} R^2\overset{\overset{O}{\|}}{C}OH + E \quad (10.12)$$
$$+$$
$$R^1OH$$

which can be interpreted for the first two stages by

$$(10.13)$$

The third stage involves H_2O in an activated complex similar to that in equation 10.13 with water taking the place of R^1OH. This step 3 can be considered to be parallel and in the reverse direction to equation 10.13.

The mechanism indicates the role of the active site binding points histidine-57 and serine-195 in the catalytic process, but the overall process is more complex and is undoubtedly not fully interpreted by this reaction sequence.

11

CHAIN REACTIONS

So far we have considered reactions in which the reacting species were normally activated molecules or ions. The previous three chapters concentrated on the molecular details of catalytic processes. In some reaction mechanisms, however, another type of molecular species, known as a *free radical*, plays a predominant role.

11.1 Free Radicals

A free radical is a molecule or ion which possesses one or more unpaired electrons. In addition, it is usual to restrict the term 'free radical' to those substances in which the unpaired electron gives rise to extraordinary chemical reactivity. The following are typical examples of free radicals. The unpaired electron is denoted by a dot.

H·	atomic hydrogen
HO·	hydroxyl
CH_3·	methyl
CH_3CO·	acetyl

Free radicals cannot be prepared in the pure state at high concentrations because of their tendency to recombine to form a normal molecule. Molecules like oxygen, nitric oxide and nitrogen dioxide contain unpaired electrons, but are not usually classified as free radicals since they are stable in the pure state, and are not abnormally reactive.

From a kinetic viewpoint the important features of free radicals are:

(a) their high reactivity can bring about reactions which cannot be readily accomplished by stable molecules;

(b) in the reaction of a free radical with an ordinary molecule one or more free radicals are among the products.

The reason for (b) is that a normal molecule contains an even number of electrons, whereas a free radical almost always contains an odd number of electrons. An odd number plus an even number always gives an odd number. Hence when a free radical reacts with a normal molecule, the products must contain an odd number of electrons, i.e. a free radical is formed. Atomic oxygen, however, has an even number of electrons and so can react with a normal molecule, e.g. hydrogen, to form two free radicals (see equation 11.14), since the sum of two odd numbers is an even number.

11.2 Types of Chain Processes

A *chain reaction* consists of a series of consecutive reactions in which a product molecule of one stage becomes a reactant in the next stage. This molecule is known as a *chain carrier*, of which the most common type is a free radical. Chain reactions are further classified as *linear* or *branched* depending on whether one or more chain carriers are formed for each chain carrier that reacts.

It might at this point be noted that branched and linear chain reactions occur as a result of nuclear fission, e.g. in atomic piles and in atomic bombs. These, however, are nuclear and not chemical processes and are not within the scope of this book.

11.3 Paneth's Lead-mirror Technique

The idea that chemical reactions can occur by chain mechanisms was first suggested by Bodenstein in 1913. This was verified by Paneth in 1929, who showed that free radicals exist as intermediates in the thermal decomposition of organic substances. As will be seen later this is strong evidence for a chain mechanism.

A schematic diagram of Paneth's apparatus is shown in *Figure 11.1*. A movable electric furnace, maintaining a temperature of 450 °C, was initially at position A. On passing the nitrogen saturated with tetramethyl-lead through the furnace a lead mirror was deposited at A. When the furnace was moved to position B a new lead mirror was formed at B while the lead mirror at A gradually disappeared. It was found that the mirror at A could be removed when B was up to 30 cm away from A. Cooling the tube between A and B did not affect the removal of the mirror.

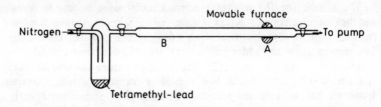

Figure 11.1 *Metallic mirror technique*

The explanation of these results is that, on first heating, the reaction

$$Pb(CH_3)_4 \rightarrow Pb + 4CH_3 \cdot$$

occurs at A. The resultant methyl radicals in the gas stream are then capable of reacting with lead at low temperatures to reform tetramethyl-lead. Thus when the furnace is moved to B the methyl radicals formed are carried to A and remove the lead mirror previously deposited there.

It was subsequently found that if organic vapours, e.g. ethanal, were passed through the furnace the resultant gases could also remove metallic mirrors and by condensing the vapours in a cold trap, metal alkyls could be isolated. The explanation in the case of ethanal is that the first stage of the decomposition is

$$CH_3 CHO \rightarrow CH_3 \cdot + CHO \cdot$$

which produces the free radicals that remove the metallic mirror.

11.4 The Hydrogen–Chlorine Reaction in Ultraviolet Light

The role of free radicals in promoting chain reactions is best seen by considering particular examples. One of the simplest is the reaction of hydrogen with chlorine. In the dark, under ordinary laboratory conditions, the reaction is very slow, but in sunlight reaction occurs quite rapidly according to the overall equation

$$H_2 + Cl_2 \rightarrow 2HCl$$

From the low rate of reaction in the dark it may be concluded that hydrogen and chlorine *molecules* do not readily react together. The mechanism suggested by Nernst (1918) is

Initiation $\qquad\qquad Cl_2 \xrightarrow[\text{light}]{\text{ultraviolet}} Cl\cdot + Cl\cdot$ $\qquad\qquad$ (11.1)

Propagation $\begin{cases} Cl\cdot + H_2 \longrightarrow HCl + H\cdot & (11.2) \\[2mm] H\cdot + Cl_2 \longrightarrow HCl + Cl\cdot & (11.3) \end{cases}$

The $Cl\cdot$ formed in equation 11.3 can then react with another hydrogen molecule, and so a continuous chain of reaction can be brought about by one initial free radical. This chain will continue until either
(a) all the hydrogen or all the chlorine is used up, or
(b) free radicals are destroyed by recombination, e.g.

Termination $\begin{cases} H\cdot + H\cdot \longrightarrow H_2 & (11.4) \\[2mm] H\cdot + Cl\cdot \longrightarrow HCl & (11.5) \\[2mm] Cl\cdot + Cl\cdot \longrightarrow Cl_2 & (11.6) \end{cases}$

Equations 11.1–11.6 illustrate the three essential processes of any chain reaction. These are:

Initiation, in which free radicals are formed from normal molecules, by chemical reaction, thermal decomposition or absorption of radiation (as in equation 11.1).

Propagation, in which reactant molecules are converted into product molecules without any change in the total number of free radicals. There may, however, be a change in the type of free radical, as in equations 11.2 and 11.3.

Termination, in which the radicals recombine to give normal molecules (as in equations 11.4, 11.5 and 11.6).

At first sight it might be expected that termination reactions would occur easily as the free radicals could reform stable molecules. This is not so, because radical recombination reactions are highly exothermic. Unless a third body removes some of the liberated energy the highly excited molecule formed in the reaction is unstable. For example, the reaction $H_2 \rightarrow H\cdot + H\cdot$ is highly endothermic (452 kJ mol^{-1}). Conversely, if two hydrogen atoms recombine, equivalent heat will be liberated. Suppose now that two hydrogen atoms recombine far away from other molecules or the walls of the vessel. The molecule formed will possess this amount of energy which is precisely the energy of dissociation. It will, therefore, be unstable and will be able to redissociate into atoms at its first vibration.

It may be concluded from this argument that radical recombinations will only occur if a third body is present to take up some of the energy liberated in the recombination. This third body is usually an atom in the surface of the walls of the vessel or another molecule in the gas phase. The latter possibility requires a three-body collision which can be shown by the kinetic theory of gases to be improbable at low pressures. Hence free radicals in the gas phase can have a life far longer than the time between collisions.

The necessity of the presence of a third body in radical recombination reactions is sometimes referred to as the *chaperon effect*. A chaperon is a third person who ensures that a highly excited meeting between two young persons of the opposite sex does not result in a temporary and unstable association. Instead the chaperon brings the couple to a stable (matrimonial) union with reduced degrees of freedom!

Proof of the chain nature of the reaction between hydrogen and chlorine comes from two other sources. First, by measuring the quantum yield, it was found that 10^4-10^6 molecules of HCl are formed for each photon absorbed. The role of light in this reaction is clearly that of an initiator leading to a chain reaction

$$Cl_2 + \text{ultraviolet light} \rightarrow Cl \cdot + Cl \cdot$$

The resultant chlorine atoms start off chains as in equations 11.2 and 11.3.

11.5   The Hydrogen–Chlorine Reaction: Polanyi's Techniq

The second method is due to M. Polanyi, who used the apparatus illustrated in *Figure 11.2*. Hydrogen, which had been passed over molten sodium, was allowed to mix with chlorine and the amounts of sodium chloride and hydrogen chloride formed were measured. It was found that

$$\frac{\text{moles of HCl formed}}{\text{moles of NaCl formed}} \approx 10\ 000$$

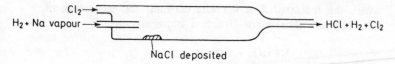

Figure 11.2   Polanyi's technique

This is interpreted by the mechanism

$$Na + Cl_2 \rightarrow NaCl + Cl\cdot$$

$$Cl\cdot + H_2 \rightarrow HCl + H\cdot$$

$$H\cdot + Cl_2 \rightarrow HCl + Cl\cdot$$

Thus each sodium atom forms one free radical which initiates a chain reaction.

11.6   Thermal Decomposition of Diethyl Ether and of Ethanal

The kinetics of chain reactions are necessarily complex but in some cases a simple order of reaction may result.   An example is the thermal decomposition of diethyl ether

$$C_2H_5OC_2H_5 \rightarrow C_2H_6 + CH_3CHO$$

which obeys first-order kinetics.   The mechanism proposed by Hinshelwood for this reaction has four stages.   It involves two free radicals in a complex process which is not discussed here. However, the thermal decomposition of ethanal has a comparatively simple mechanism which can be shown to lead to an order of 1.5, as found by experiment.   The mechanism is

Initiation $\qquad CH_3CHO \overset{k_1}{\rightarrow} CH_3\cdot + CHO\cdot \qquad\qquad$ (11.7)

Propagation $\qquad CH_3CHO + CH_3\cdot \overset{k_2}{\rightarrow} CH_4 + CO + CH_3\cdot \qquad$ (11.8)

Termination $\qquad \begin{cases} 2CH_3\cdot \overset{k_3}{\rightarrow} C_2H_6 \\[2mm] 2CHO\cdot \overset{k_4}{\rightarrow} 2CO + H_2 \end{cases} \qquad\qquad$ (11.9)

where the indicated k is the rate constant for each stage of the reaction. The overall reaction is mainly

$$CH_3CHO \rightarrow CH_4 + CO$$

although traces of C_2H_6 and H_2 are formed.

If the reaction proceeds at a steady rate it may be assumed that the rate of formation of methyl radicals in equation 11.7 is equal to the rate of their destruction in equation 11.9, since the concentration of $CH_3\cdot$ is unaffected by equation 11.8. This follows from the stationary state hypothesis (see p. 71).

Rate of formation $= k_1[CH_3CHO]$

$\qquad\qquad\qquad = k_3[CH_3\cdot]^2$ = rate of destruction $\qquad$ (11.10)

whence $[CH_3\cdot] = \left(\dfrac{k_1}{k_3}[CH_3CHO]\right)^{1/2}$ (by rearranging 11.10)

Since the rate of formation of product is governed by equation 11.8,

overall rate $= k_2[CH_3CHO][CH_3\cdot]$

$\qquad\qquad = k_2\left(\dfrac{k_1}{k_3}\right)^{1/2}[CH_3CHO]^{3/2}$ $\qquad\qquad\qquad$ (11.11)

The reaction is therefore of order 3/2.

Free radicals have been detected in the thermal decompositions of both diethyl ether and ethanal. Since the observed orders of reaction agree with those predicted by free radical mechanisms, it may be concluded that both these decompositions are chain reactions.

11.7 Polymerization Reactions

Chain reactions can also occur in the liquid phase and are the basis of many industrial processes for forming polymers. A common example is the polymerization of molecules of the form $RCH=CH_2$ according to the following equations, where R is a substituent group in the ethene molecule and $X\cdot$ is a free radical:

$$X\cdot + RCH=CH_2 \rightarrow RXCHCH_2\cdot$$
$$RXCHCH_2\cdot + RCH=CH_2 \rightarrow RXCHCH_2RCHCH_2\cdot$$

This process can be repeated n times to give

$$RXCHCH_2(RCHCH_2)_n \cdot$$

Eventually the chain is terminated by radical recombination. Some examples of industrially important polymers are shown in *Table 11.1*, which deals with the polymerization of ethene molecules, mono-substituted by the group R. Perspex and Teflon are

Table 11.1

R	Polymer
H	Polyethene
CH_3	Polypropene
Cl	Poly(vinyl chloride) (PVC)
$OCOCH_3$	Poly(vinyl acetate) (PVA)
C_6H_5	Polystyrene

also polymerized ethene derivatives, the monomers being methyl methacrylate and tetrafluoroethene, respectively.

The kinetics of such reactions are complex, but their mechanisms all involve the common features of *initiation*, *propagation* and *termination*, which are characteristic of all chain reactions.

Initiation of polymerization reactions can be brought about by irradiation with ultraviolet light or gamma rays or by chemical methods. The addition of benzoyl peroxide or Fe^{2+} and H_2O_2 (Fenton's reagent) yields free radicals according to the equations

$$C_6H_5COOOH \rightarrow C_6H_5COO \cdot + \cdot OH$$

$$Fe^{2+} + H_2O_2 \rightarrow Fe^{3+} + \cdot OH + OH^-$$

11.8 Identification of Chain Reactions

The following methods may be used to determine whether a reaction follows a chain mechanism.

11.8.1 ADDITION OF INHIBITOR

If a small amount of an added substance, such as nitric oxide or propene, produces a large decrease in the rate of reaction, a chain reaction may be suspected. The inhibitor is a molecule that can react with and remove free radicals, for example

$$CH_3 \cdot + \cdot NO \rightarrow CH_3NO \rightarrow H_2O + HCN$$

Nitric oxide is an effective inhibitor and a partial pressure of 1–2 mmHg may reduce the rate of reaction by a factor of 10–100 or even stop it altogether. The reason is that a nitric oxide molecule removes a free radical and so prevents the starting of a chain. Propene also acts as an inhibitor in gas reactions, being itself polymerized in the process. Benzoquinone is an effective inhibitor of liquid-phase polymerization reactions.

The term *'negative catalyst'* is sometimes used for inhibitors. This is not desirable as the inhibitor reacts irreversibly with the free radical in carrying out its function whereas a catalyst is still in its original chemical state at the end of the reaction.

11.8.2 METALLIC MIRROR TECHNIQUE

If the gaseous products of reaction are capable of removing metallic mirrors, by forming metal alkyls, then it may be concluded that free radicals are present and a chain mechanism is likely.

11.8.3 PHYSICAL METHODS

In some cases free radicals may be detected directly by physical methods, e.g. spectroscopy, mass spectrometry or electron spin resonance. The existence of free radicals is again evidence of a chain mechanism.

11.9 Branched Chain Reactions

Up to this point, the reactions considered in this chapter have been examples of linear chain processes. Some chain reactions involve an additional feature known as *branching*, which occurs when more than one free radical is produced for each free radical consumed. The combustion of hydrogen and hydrocarbons in air or oxygen frequently gives rise to branched chain reactions. This will be illustrated by considering the reaction of hydrogen and oxygen to form water at 500–600 °C:

Initiation $\qquad$ $W + H_2 \rightarrow H\cdot + H\cdot$ $\qquad$ (11.12)

where W refers to an inert atom on the walls of the vessel

Branching $\qquad$ $H\cdot + O_2 \rightarrow \cdot OH + O\cdot$ $\qquad$ (11.13)

Branching $\qquad$ $O\cdot + H_2 \rightarrow \cdot OH + H\cdot$ $\qquad$ (11.14)

Propagation $\qquad$ $\cdot OH + H_2 \rightarrow H_2O + H\cdot$ $\qquad$ (11.15)

Propagation $\left\{\begin{array}{l} H\cdot + O_2 \rightarrow HO_2\cdot \qquad\qquad (11.16) \\ HO_2\cdot + H_2 \rightarrow H_2O + \cdot OH \qquad (11.17) \end{array}\right.$

Termination $\qquad$ Radicals $\rightarrow$ inactive products at walls of vessel

The special feature of this mechanism is the two branching reactions 11.13 and 11.14 in which the number of free radicals doubles. This makes possible a rapid increase in the rate of reaction which will lead to explosion unless the termination processes can destroy the extra free radicals as fast as they are produced. A high temperature is necessary to bring about reaction because of the difficulty of producing hydrogen atoms from hydrogen molecules by direct heating.

On the other hand, the reaction of hydrogen and oxygen can occur explosively at room temperature on sparking. The spark produces a local 'hot spot' which produces sufficient free radicals to trigger off the branched chain.

At first sight it may not appear likely that, by starting with a few free radicals and continuously doubling their number, an explosion should be produced. The following story illustrates the effect of continuous doubling in tangible terms.

It is said that in ancient times a Pharaoh was taught to play chess and was so pleased with the game that he offered his tutor a reward. The tutor demanded one grain of wheat for the first square on the board, two for the second, four for the third, eight for the fourth and so on for all 64 squares. This seemed a modest prize and was agreed to by the Pharaoh. However, when the quantity of wheat is calculated [$(2^{64} - 1)$ grains] it comes to about a million million tons! In a gas at STP an individual molecule would suffer 64 collisions in less than a millionth of a second. If each collision resulted in branching an explosion would certainly occur.

The oxidation of hydrocarbons (of importance in internal com bustion engines) is also a branched chain process, the initial stage being

Initiation $\qquad RH + O_2 \rightarrow HO_2\cdot + R\cdot$

Propagation $\begin{cases} R\cdot + O_2 \rightarrow RO_2\cdot \\ RO_2\cdot + RH \rightarrow RO_2H + R\cdot \end{cases}$

Branching $\qquad RO_2H \begin{array}{l} \nearrow OH\cdot + \text{ organic radical} \\ \searrow \text{ molecular oxidation products} \end{array}$

The function of tetraethyl-lead as an *antiknock* in petrol is to con trol the number of free radicals by providing a catalytic surface lead oxide particles on which radicals can recombine. Many processes of combustion are also branched chain reactions.

In photochemical reactions and in radiolysis, the primary produ are free radicals. These undergo secondary reactions to form th final products. Many chain reactions can be initiated by the act of light or of ionizing radiation.

It should be noted that in some cases heterogeneous catalysis and a branched chain reaction may both be occurring. The oxi dation of naphthalene to form phthalic anhydride is a major indus trial process

$$4\tfrac{1}{2}O_2 + \text{[naphthalene]} \xrightarrow{V_2O_5} \text{[phthalic anhydride]} O + 2H_2O + 2CO_2$$

It is carried out by passing naphthalene vapour and air over a heated vanadium pentoxide catalyst. In addition to the reaction on the catalyst, a branched chain oxidation also occurs in the gas phase. If conditions are not correctly chosen this can result in a chain explosion.

The kinetics of branched chain reactions will not be dealt with in this text.

12

FAST REACTIONS

Following the development of new physicochemical techniques, the study of fast reactions is now an important part of reaction kinetics. All that will be attempted in this chapter is to high-light some of the important features of this topic. The interested reader requiring a detailed treatment should consult a specialist text such as E.F. Caldin's *Fast Reactions in Solution*.

The rates of many reactions are too fast to be measured by the conventional methods described in Chapter 3. These methods cannot deal with reactions whose half-lives are less than a second or so. Such reactions are classified as 'fast'. Typical examples of fast reactions are:

(a) many ionic reactions such as the neutralization of acids by bases;

(b) reactions of biological significance, e.g. reactions between an enzyme and a substrate; the reaction of haemoglobin with oxygen and carbon monoxide;

(c) organic substitution reactions, such as the bromination of aromatic amines and phenols;

(d) the explosive reaction of oxygen with hydrogen and hydrocarbons.

Some of these reactions were once regarded as instantaneous. Their rates can now be measured.

To put into perspective how fast these 'fast reactions' are, it is worth considering a set of hypothetical second-order one-stage reactions between two reactants each of which has an initial concentration of 1.0 mol l^{-1}.

Using equation 2.16 the half life of such a reaction is given by

$$t_{1/2} = 1/ka$$

from which result the data in *Table 12.1* are calculated.

Table 12.1 Relation of half-life to rate constant. Second-order reactions with reactants at the initial concentration of 1.0 mol l^{-1}

Half-life/s	$k/$l mol^{-1} s^{-1}	*Comments*
10^7 (about 4 months)	10^{-7}	Too slow for the conventional methods of Chapter 3
10^4 (about 3 hours)	10^{-4}	Suitable for sampling methods
10	10^{-1}	Suitable for continuous methods such as spectrophotometry and conductometry
10^{-2}	10^2	Suitable for flow methods
$<10^{-5}$	$>10^5$	Special techniques needed, e.g. relaxation methods

The fastest known reactions in aqueous solution are proton transfer reactions, such as that between hydrogen ions and hydroxide ions, the rate constant being 1.4×10^{11} l mol^{-1} s^{-1} at 25 °C. The methods used in the study of fast reactions are:

(a) modification of classical methods;
(b) flow methods;
(c) relaxation methods;
(d) other special techniques.

These methods are discussed in the following sections.

12.1 Modification of Classical Methods

If the reactant concentrations can be sufficiently decreased, the half-life may be increased until it is in the classically measurable range. It is essential that an analytical method which operates accurately in dilute solution is available. An example is the potentiometric study of the bromination of phenols, described in Section 13.6. The concentration of bromine is determined from the redox potential of a bright platinum electrode immersed in the solution. Bromine concentrations as low as 10^{-8} mol l^{-1} can be measured in this way.

Another method is to use stationary state conditions so that the 'fast reaction' being investigated is limited to the slower rate of a reactant-generating process. For instance, in constant current coulometry a reactant such as bromine or iodine can be generated at a steady and measurable rate at an electrode immersed in a halide solution

$$2X^- \rightarrow X_2 + 2e^-$$

where X = Br or I.

A fast reacting substrate such as ascorbic acid (vitamin C) or arsenic(III), if present in the solution, will react at a rate imposed by the electrical generating process. Conventional analytical techniques capable of determining the low concentration of stationary-state reactant are used. Potentiometry has already been mentioned as suitable in this context, and amperometric techniques have also proved to be effective.

In enzyme kinetic studies the stationary-state approach has made important contributions to the study of reaction mechanisms.

12.2 Flow Methods

The simplest flow method is illustrated schematically in *Figure 12.1(a)*. Reagents are pumped continuously through the reaction vessel. Partial reaction occurs and a steady state is set up. A mixture of reactants and products flows from the reaction vessel. The composition of the outflow must be analysed either instrumentally or by 'freezing' the mixture and then analysing it by a conventional method.

Suppose the reaction involved is a second-order one

$$A + B \overset{k}{\rightarrow} C$$

Let a, b and c be the concentrations of A, B and C when a steady state has been set up in the thoroughly stirred reaction vessel. It is then assumed that the rate at which C is generated in the reaction vessel (volume V litres) equals the rate at which C is removed from the reaction vessel via the outlet. If the rate of flow in the outlet is R (1 s^{-1}) then the mass balance requires

$$kabV = Rc \qquad (12.1)$$

from which the rate constant can be found. This method is particularly useful in studying intermediates and can be used for values of k up to 10 1 mol^{-1} s^{-1}.

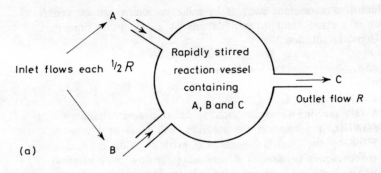

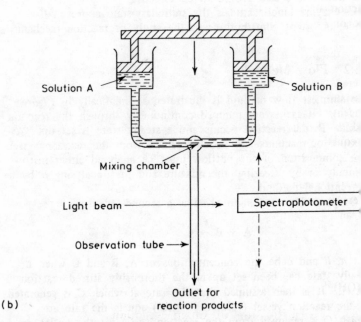

Figure 12.1 Flow methods (schematic). (a) Principle of flow method; (b) Hartridge and Roughton apparatus

The limiting factor with this type of apparatus is the time of mixing, which must be much less than the half-life of the react being measured if significant results are to be obtained.

Hartridge and Roughton (1923) set up a continuous flow app atus with a low-volume mixing chamber. A schematic diagram their apparatus is shown in Figure 12.1(b). It was possible to reduce the time of mixing to 10^{-3} s by the use of a specially designed mixing chamber with tangentially positioned jets.

The reactants are forced under pressure through the mixing chamber from where they flow at high speeds (up to 10 m s^{-1}) through the observation tube. Distance along the observation tube is equivalent to the time elapsed after mixing the reagents. A distance of 1 cm along the tube corresponds to 10^{-3} s if the flow rate is constant at 10 m s^{-1}. The concentrations in the observation tube are determined by a spectrophotometric method at various positions (i.e. various times after mixing). In this way the rates of reactions with half-lives of a few milliseconds can be followed.

One of the first reactions studied by this method was that between ferric and thiocyanate ions in aqueous solution. (This reaction appears to be instantaneous.) It was shown that the rate of reaction varies with the acidity:

$$\text{rate} = k[Fe^{3+}][CNS^-][1 + a/[H^+]] \tag{12.2}$$

where k is the second-order rate constant, and a is an empirical constant related to the dependence of the reaction rate on the concentration of hydrogen ions. At 25 °C in 0.1 M perchloric acid, $k = 127$ l mol^{-1} s^{-1}. The kinetic explanation of the form of equation 12.2 is in terms of the following equilibrium

$$Fe^{3+} + H_2O \overset{K}{\rightleftharpoons} Fe(OH)^{2+} + H^+ \tag{12.3}$$

where K is the equilibrium constant. From the usual expression for an equilibrium constant

$$[Fe(OH)^{2+}] = K[Fe^{3+}]/[H^+] \tag{12.4}$$

the observed kinetic behaviour can be explained if both Fe^{3+} and $Fe(OH)^{2+}$ take part in the reaction, i.e.

$$\text{rate} = k_1[Fe^{3+}][CNS^-] + k_2[Fe(OH)^{2+}][CNS^-] \tag{12.5}$$

Putting 12.4 into 12.5 and rearranging

$$\text{rate} = k_1[Fe^{3+}][CNS^-](1 + k_2K/k_1[H^+]) \tag{12.6}$$

which is of the same form as equation 12.2, the empirical constant a being replaced by k_2K/k_1. These kinetic measurements, therefore, indicate that $Fe(OH)^{2+}$ takes part in the reaction.

The flow method has also been used to confirm the Michaelis–Menten mechanism of enzyme substrate reaction in the case of the reaction of catalase and H_2O_2.

12.3   Relaxation Methods

There is no method by which two separate solutions can be thoroughly mixed in less than 10^{-3} s.   Reactions occurring in less than this time can only be studied by a method that avoids mixing separate solutions.

One way of doing this is to subject a reversible reaction at equilibrium to a sudden change in conditions (e.g. temperature, pressure or electric field strength).   The reaction is no longer in equilibrium, but moves rapidly to its new position of equilibrium. This movement to equilibrium (known as *relaxation*) can be measured by a 'fast' physical method.   Kinetic studies based on the measurement of the rate of attainment of equilibrium are known as *relaxation methods.*

The analytical method used (such as conductometry or spectro-photometry) must be capable of immediately converting the fraction of reaction into an electrical signal which can be fed to the Y-plates of a cathode-ray tube.   The sudden change on the system which produces the new equilibrium conditions is made to trigger the time base of the cathode-ray tube.   In this way results similar to those shown in *Figure 12.2* can be obtained.   Times between 1 and 10^{-6} s can be conveniently handled in this way.

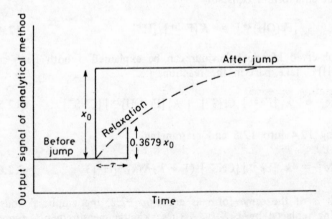

Figure 12.2   Relaxation process showing relaxation time

Suppose the equilibrium being studied is the reversible first-order reaction

$$A \underset{k_{-1}}{\overset{k_1}{\rightleftharpoons}} B \qquad (12.7)$$

Now suppose the equilibrium is suddenly displaced to the right. Let the original and final concentrations of A and B be a_0, a_e and b_0, b_e respectively and let a and b denote the concentrations at time t. If the initial displacement from the new equilibrium position is x_0, and x is the displacement at time t, then

$$x = (a - a_e) = (b_e - b) \qquad (12.8)$$

The net rate of attainment of the new equilibrium position, i.e. the *rate of relaxation*, is

$$-dx/dt = k_1 a - k_{-1} b \qquad (12.9)$$

Substituting 12.8 into 12.9

$$-dx/dt = k_1(x + a_e) - k_{-1}(b_e - x) \qquad (12.10)$$

which on collecting terms gives

$$-dx/dt = (k_1 a_e - k_{-1} b_e) + (k_1 + k_{-1})x \qquad (12.11)$$

Now the first bracket on the right-hand side of equation 12.11 is zero since at equilibrium $k_1 a_e = k_{-1} b_e$. Hence

$$-dx/dt = (k_1 + k_{-1})x \qquad (12.12)$$

Rearranging 12.12

$$-dx/x = (k_1 + k_{-1})dt \qquad (12.13)$$

Integrating equation 12.13 using M21

$$-\ln x = (k_1 + k_{-1})t + \text{constant} \qquad (12.14)$$

The integration constant is evaluated by using the condition that when $t = 0$, $x = x_0$. Hence the constant in equation 12.14 is $-\ln x_0$. Finally

$$\boxed{\ln(x_0/x) = (k_1 + k_{-1})t} \qquad (12.15)$$

Equation 12.15 illustrates the important principle that the rate of relaxation depends on *two* rate constants. These can, however, be individually evaluated if the equilibrium constant K is also known, for

$$K = k_1/k_{-1} \qquad (12.16)$$

The rates of relaxation processes are defined by a *relaxation time* (τ) which is the time at which $\ln(x_0/x) = 1.00$. This is equivalent to $x/x_0 = 1/e = 1/2.718 = 0.3679$, i.e. at the relaxation time the system has gone 36.79% of the way to the new equilibrium position.

For the reaction shown in equation 12.7 the relaxation time given by using equation 12.15 is

$$\tau = 1/(k_1 + k_{-1}) \qquad (12.17)$$

In general, the reactions involved in the relaxation process may be of higher order than unity. The mathematical treatment of relaxation is then more difficult, but still in accordance with the principles outlined above.

12.4 Experimental Techniques in Relaxation Methods

Many types of relaxation methods have been used in studying the kinetics of fast reactions. Two of the simplest to understand are

(i) the temperature-jump method;
(ii) the pressure-jump method.

12.4.1 THE TEMPERATURE JUMP METHOD

In the temperature jump method an increase of temperature of up to $10\,^{\circ}C$ can be produced in 10^{-6} s by the discharge of a high voltage condenser (100 kV) through a small quantity of solution. The increase in temperature changes the equilibrium constant in accordance with the van't Hoff isochore (p. 36).

The temperature-jump method has been used to study the reaction $H^+ + OH^- \rightarrow H_2O$. At $25\,^{\circ}C$ the relaxation time is about 35×10^{-6} s.

12.4.2 THE PRESSURE JUMP METHOD

In the pressure jump method the reaction mixture is enclosed in a pressure vessel sealed with a thin metal disc, and subject to a pressure of about 5×10^6 Pa (50 atm). This pressure can be reduced to the atmospheric value in about 10^{-4} s by puncturing the metallic disc. The decrease in pressure changes the equilibrium constant in accordance with equation 5.20 (p. 63).

In both cases the relaxation process is followed by a fast physical method of analysis.

12.5 Other Special Techniques

Many instrumental techniques have been introduced in recent years to the study of fast reactions. The following examples illustrate the range of methods used.

12.5.1 FLASH PHOTOLYSIS

A powerful flash of light of duration about 10^{-5} s and energy about 10^5 J is produced by discharging high-voltage capacitors through an inert gas at low pressure. The flash produces photochemical reactions in an adjacent reaction vessel, usually leading to relatively large free radical concentrations. The rate of decomposition of these free radicals can be followed by a fast optical method. As an example of the application of this method, a mixture of chlorine and oxygen produces chlorine monoxide (ClO) after photolysis. This is an unstable free radical which decomposes into its elements in a second-order reaction having a rate constant of about 10^8 1 mol^{-1} s^{-1} at 25°C. Measurable concentrations of ClO cannot be obtained by other methods.

12.5.2 FLUORESCENCE METHODS

Molecular fluorescence can be described as a three-stage process. Firstly, a photon of electromagnetic radiation (usually ultraviolet or visible light) is absorbed by a molecule to form an electronically excited metastable state. The excited state is usually one of high vibrational energy.

In the second stage, most of the excess vibrational energy is dissipated to adjacent molecules. If the fluorescent species is in solution this will be to the surrounding solvent molecules. In the third stage, fluorescence occurs when the excited state reverts to the ground state with the emission of a photon of radiation. The emitted radiation has in general a lower energy and hence a longer wavelength than that of the incident photon.

The fluorescence spectrum depends on energy level differences in the molecule being excited, and is independent of the exciting radiation. For example, molecules of rhodamine B will show the same red fluorescence when excited by either ultraviolet or visible radiation.

A substance that reacts with the electronically excited metastable molecule and removes its excess energy, so preventing the re-emission of radiation, is said to *quench* the fluorescence. As the time between excitation and re-emission of radiation is of the order of 10^{-8} s, the de-activating reaction must be very fast if effective quenching is to occur.

The reduction in the intensity of the fluorescent radiation due to the quenching agent can be measured and the rate constant for the quenching reaction can then be calculated by assuming a reaction order for the quenching process and from a knowledge of the concentration of the quenching agent.

The fluorescence of chlorophyll is strongly quenched by oxygen. This very fast reaction has been extensively studied by fluorescence spectroscopy and a rate constant of the order of 10^{10} l mol^{-1} s^{-1} has been obtained.

Many co-enzymes give rise to intense fluorescence and so can easily be detected in very low concentrations. Accordingly, many rapid enzyme reactions which involve co-enzymes are monitored with comparative ease by fluorescence methods.

12.5.3 NUCLEAR MAGNETIC RESONANCE (NMR)

A study of the width and shape of the NMR absorption lines under high resolution gives information about the average environment of the atom whose absorption is being measured.

In general, lines are broadened and fine structure is lost if rapid exchange reactions are taking place. The exchange of protons between water, ammonia and ammonium ions is an example of this. Similar techniques can be used in the study of free radicals by electron spin resonance (ESR).

12.5.4 POLAROGRAPHY

This method is based on measuring the current produced by applying a potential difference across a dropping mercury electrode and a reference electrode. By means of the theoretically derived Ilkovic equation it is possible to predict the current at the top of a polarographic wave.

Most polarographic waves are accurately described by the Ilkovic equation, and are said to be *diffusion controlled.* In certain cases the current is much less than the predicted value as the reducible species is being generated from other molecules and the rate of this reaction determines the extent of reaction on the dropping mercury surface. Such systems are said to be *kinetically controlled.*

An example is provided by a solution of methanal in water. This is extensively hydrated

$$HCHO + H_2O \rightleftharpoons HOCH_2OH$$

and only the free methanal is reducible at the mercury surface. Analysis of the kinetic data enables the rate of breakdown of the methylene diglycol to be calculated.

12.5.5 MOLECULAR BEAMS

At very low pressures (less than 10^{-5} mmHg) the mean free path of gas molecules is much greater than one metre. Thus it is possible to create a flow of effectively non-colliding molecules. Such a flow is termed a *molecular beam.* By arranging for two beams to intersect, the occasional encounters that result will bring about scattering of the colliding molecules. The intensity of scattering in various directions is determined with the aid of a movable detector.

From experiments on crossed molecular beams, the dynamics of bimolecular collisions can be studied. Reactions most amenable to this form of investigation are of the type

$$RX + M \rightarrow R + MX$$

where R = alkyl, halogen or hydrogen

X = halogen and

M = alkali metal

The *reactive cross section*, which indicates the probability of a collision producing chemical reaction, is high for these species.

ANALYSIS OF EXPERIMENTAL RESULTS

13.1 Calculation of a First-order Rate Constant from the Differential Rate Law

The method is illustrated using data taken from the decomposition of benzenediazonium chloride[*] in water at 40 °C:

$$C_6 H_5 N_2 Cl + H_2 O \rightarrow C_6 H_5 OH + HCl + N_2$$

The reaction was followed by measuring the increase in pressure, due to the evolution of nitrogen, by means of a xylene manometer. Consequently, the amount of reaction is expressed in terms of centimetres of xylene. The experimental results are shown in *Table 13.1*. p represents the increase in pressure due to the evolved nitrogen; p_∞ is the final pressure after infinite time.

Table 13.1

$t/$ min	$p/$ cm xylene	$(p_\infty - p)/$ cm xylene	$\log\left\{\dfrac{(p_\infty - p)/}{\text{cm xylene}}\right\}$
0	0	22.62	1.3545
2	1.07	21.55	1.3334
4	2.15	20.47	1.3111
8	4.14	18.48	1.2667
12	5.89	16.73	1.2235
16	7.52	15.10	1.1790
20	9.00	13.62	1.1342
24	10.32	12.30	1.0899
28	11.52	11.10	1.0453
35	13.33	9.29	0.9680
40	14.47	8.15	0.9112
45	15.42	7.20	0.8573
50	16.28	6.34	0.8021
60	17.74	4.88	0.6884
68	18.60	4.02	0.6042
'infinity'	22.62	–	–

[*]Moelwyn-Hughes, E.A. and Johnson, P., *Trans. Faraday Soc.*, **36**, 950 (1940)

The rate law for a first-order reaction (equation 2.6) is

$$\frac{\mathrm{d}x}{\mathrm{d}t} = k(a - x)$$

k may be calculated from this equation by substituting into it numerical values of $\mathrm{d}x/\mathrm{d}t$ and $(a - x)$ at a given time. In this experiment the fraction of reaction is proportional to the increase in pressure p; the initial concentration a is proportional to p_∞. Therefore the concentration remaining at time t, i.e. $(a - x)$, is proportional to $(p_\infty - p)$. For a first-order reaction, any units may be used to measure concentration since, from equation 2.10, k depends on a *ratio* of two concentrations, the ratio being dimensionless. This justifies the use of cm xylene as the unit of concentration in *Figure 13.1*.

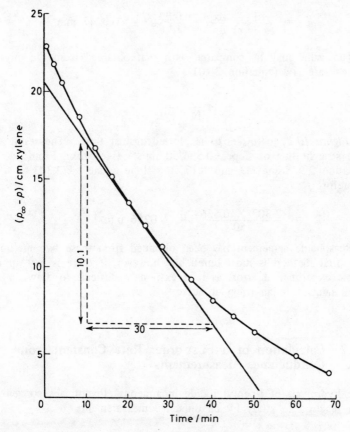

Figure 13.1 Decomposition of benzenediazonium chloride at 40.0 °C (original results)

Using the data in *Table 13.1*, a graph of $(a - x)$ against t has been plotted in *Figure 13.1*, and a tangent has been drawn to the curve at $t = 20$ min.

$$\text{Slope of graph} = \frac{d(a - x)}{dt} = \frac{-dx}{dt}$$

Using M18, slope of graph = slope of tangent

$$= -\frac{10.1}{30} = -0.337 \text{ cm xylene min}^{-1}$$

At this time $(a - x)$ corresponds to 13.62 cm xylene. Substituting in equation 2.6

$$k = \frac{1}{(a - x)} \frac{dx}{dt} = \frac{0.337}{13.62} = 0.0247 \text{ min}^{-1}$$

This value may be compared with that obtained from the integrated rate law (equation 2.10)

$$kt = \ln \frac{a}{(a - x)}$$

In *Figure 13.2*, $\log(p_\infty - p)$ is plotted against time. The result is a straight line of slope $-0.556/50$ min^{-1} (see M2). From equation 2.9, using M3 and M17, the slope equals $-k/2.303$. Therefore

$$k = \frac{2.303 \times 0.556}{50} = 0.0256 \text{ min}^{-1}$$

Reasonable agreement has been obtained from these two methods. The first method is most unreliable, however, if there is any appreciable experimental error, as it is extremely difficult to draw a good tangent to an irregular curve.

13.2 Calculation of a First-order Rate Constant from Conductance Measurements

The hydrolysis of t-pentyl iodide in aqueous ethanol is a convenient reaction to study by conductance measurements

$$\text{t-C}_5\text{H}_{11}\text{I} + \text{H}_2\text{O} \rightarrow \text{t-C}_5\text{H}_{11}\text{OH} + \text{H}^+ + \text{I}^-$$

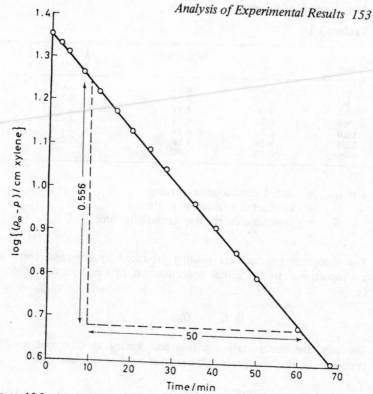

Figure 13.2 Decomposition of benzenediazonium chloride at 40.0 °C (first-order plot)

In this reaction the concentration of ions increases with time. In dilute solutions, the conductance is proportional to the concentration of ions; consequently the increase in conductance is proportional to the amount of reaction.

The experiment is carried out by preparing an approximately 0.02 molar solution of t-pentyl iodide in aqueous ethanol, the conductance of which is measured at various times using any convenient conductance cell. Since first-order rate constants depend on a *ratio* of concentrations (equation 2.10), it is not necessary to know the cell constant, nor to include the range of the scale used provided that all the measurements are related to the same scale.

The results obtained in an experiment using 80% aqueous ethanol as solvent at 25 °C are shown in *Table 13.2*.

The integrated rate law for a first-order reaction (equation 2.10) is

$$kt = \ln \frac{a}{a - x}$$

Table 13.2

$t/$ min	$G/$ S	$(G_\infty - G)/$ S	$\log\left\{\dfrac{(G_\infty - G)/}{S}\right\}$
0	0.39	10.11	1.005
1.5	1.78	8.72	0.941
4.5	4.09	6.41	0.807
9.0	6.32	4.18	0.621
16.0	8.36	2.14	0.330
22.0	9.34	1.16	0.065
'infinity'	10.50	0.00	—

Let G_0 = initial conductance reading
$\quad G$ = conductance reading at time t
$\quad G_\infty$ = conductance reading at infinite time

The change in conductance reading produced by complete reac is proportional to the initial concentration of the t-pentyl iodi i.e.

$$a \propto G_\infty - G_0$$

Similarly, the concentration x that has reacted at time t is pr portional to the change in conductance reading at that time

$$x \propto G - G_0$$

By subtraction

$$a - x \propto G_\infty - G$$

From equation 2.9

$$\ln a - \ln(a - x) = kt$$

From M3, a graph of $\ln(a - x)$ against t will give a straight of slope $-k$. Using M16, since $(G_\infty - G)$ is proportional to $(a - x)$, a graph of $\ln(G_\infty - G)$ against t gives the same slop namely $-k$.

In *Figure 13.3* a graph is shown of $\log(G_\infty - G)$ against t. Since common logarithms are used the slope (from M17) is $-k/2.303$. From *Figure 13.3* the slope is $-0.83/20 = -0.0415$ min^{-1} (see M2). Hence $k = 2.303 \times 0.0415 = 0.095$ min^{-1}.

It may be noted that at no stage in the calculation is the actual value of concentration of reactant required. All that is

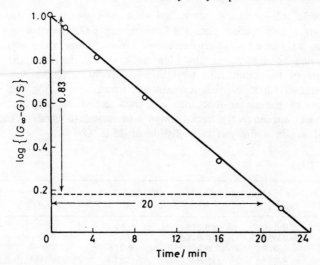

Figure 13.3   Hydrolysis of t-pentyl iodide at 25.0 °C

needed is a measurement that is proportional to concentration. The above method may therefore be used in cases where the reaction rate has been followed by measuring, for example, the pressure of evolved gas, the change in optical rotation or the absorbance of a solution.

13.3   Calculation of a Second-order Rate Constant with Equal Initial Concentrations of Reactants

The evaluation of the second-order rate constant using the method on p. 17 is shown for the reaction* of ethyl *p*-nitrobenzoate with sodium hydroxide in aqueous acetone (40% by volume) at 15.2 °C:

$$p\text{-}O_2NC_6H_4CO_2C_2H_5 \rightarrow$$

$$p\text{-}O_2NC_6H_4CO_2Na + C_2H_5OH$$

The reaction is rapid (half-life 4.1 minutes) and so a special technique was used to reduce timing errors on mixing the reactants. The method employed was to seal 5.00 ml of 0.1 M sodium

*Newling, W.B.S. and Hinshelwood, C.N., *J. Chem. Soc.*, 1358 (1936)

hydroxide solution in a thin-walled glass tube which was placed in a larger and thicker glass tube containing 5.00 ml of a 0.1 M solution of the ethyl *p*-nitrobenzoate. When temperature equilibrium had been achieved, the inner tube was broken and the contents of the outer tube were stirred rapidly. At the end of the required time, the reaction was quenched by pouring the contents of the larger tube into a known excess of standard acid, which was subsequently back-titrated with standard alkali. The rate of acidic hydrolysis is negligible at 15.2 °C.

Table 13.3

$t/$ s	$(a - x)/$ mol l^{-1}	$x/$ mol l^{-1}	$\dfrac{x}{a - x}$	$k/$ 1 mol^{-1} s^{-1}
0	0.0500	0.0000	0.000	–
120	0.0335	0.0165	0.492	0.0819
180	0.0291	0.0209	0.718	0.0796
240	0.0256	0.0244	0.953	0.0794
330	0.0209	0.0291	1.381	0.0839
530	0.0155	0.0345	2.226	0.0840
600	0.0148	0.0352	2.378	0.0792

The experimental results are shown in *Table 13.3* in which *a* represents the initial concentration of NaOH. The concentration $(a - x)$ of sodium hydroxide shown in *Table 13.3* will also be the concentration of the ethyl *p*-nitrobenzoate, since the initial concentrations are equal.

The differential rate law 2.13 is

$$\frac{dx}{dt} = k(a - x)^2$$

The corresponding integrated form 2.15 is

$$kt = \frac{x}{a(a - x)}$$

In *Figure 13.4*, $x/(a - x)$ is plotted against t. The slope of the graph is $2.06/500 = 4.12 \times 10^{-3}$ s^{-1} (see M2).

From 2.15 the slope is ak, and since $a = 0.0500$ mol l^{-1}

$$k = \frac{4.12 \times 10^{-3}}{0.0500} = 0.0824 \text{ 1 mol}^{-1} \text{ s}^{-1}$$

The values of k found by direct substitution in equation 2.15 are also shown in *Table 13.3*. The average value is 0.0813 1 mol^{-1} s^{-1}, in good agreement with the graphical value.

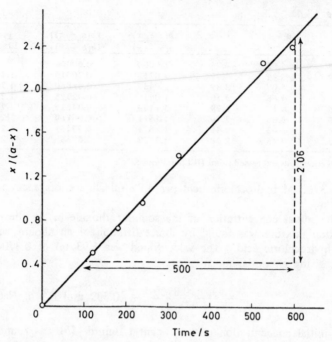

Figure 13.4   Hydrolysis of ethyl p-nitrobenzoate at 15.2 °C

13.4   Calculation of a Second-order Rate Constant when the Reactants are at Different Initial Concentrations

The method is illustrated by the reaction[*] of n-pentyl fluoride and sodium ethoxide:

$$n\text{-}C_5H_{11}F + NaOC_2H_5 \rightarrow NaF + n\text{-}C_5H_{11}OC_2H_5$$

at 120.0 °C in 99.9% ethanol. Since ethanol boils at 78 °C the sealed-tube technique must be used, and the high temperature of the experiment requires a correction for solvent expansion to be made. The experimental results are shown in *Table 13.4*. The reaction was followed by adding excess hydrochloric acid (0.3700 M) and back-titrating with standard sodium hydroxide solution. Each sealed tube contained 4.07 ml of reaction mixture (at 20 °C) and so concentrations are conveniently measured in ml

[*]Latham, J.L., *Ph.D. Thesis,* University of London, 46 (1951)

Table 13.4*

$t/$ 10^4 s	$a - x$	$b - x$	$\dfrac{b(a - x)}{a(b - x)}$	$\log\left\{\dfrac{b(a - x)}{a(b - x)}\right\}$ 10^{-5}	$k/$ 1 m
0.00	5.93	4.75	1.0000	0.00000	
1.59	5.47	4.29	1.0213	0.00915	1.2
3.30	5.07	3.89	1.0437	0.01860	1.2
5.14	4.68	3.50	1.0710	0.02978	1.2
8.16	4.17	2.99	1.1174	0.04817	1.2
10.25	3.88	2.70	1.1511	0.06110	1.2
13.04	3.63	2.45	1.1871	0.07451	1.2
17.33	3.26	2.08	1.2552	0.09868	1.2

*Concentrations expressed in ml HCl per aliquot.

of 0.3700 M hydrochloric acid per 4.07 ml aliquot of reaction
mixture.

The initial concentration of the sodium ethoxide (a) in the
reaction mixture was found by direct titration of an aliquot w
the hydrochloric acid. The value found was 5.93 ml of 0.370
HCl, giving an initial molarity of

$$\frac{5.93 \times 0.3700}{4.07} = 0.539 \text{ mol l}^{-1} \text{ of Na}$$

The initial concentration of the n-pentyl fluoride (b) was foun
by weighing a sample into a 100 ml volumetric flask and dilu
to the mark with sodium ethoxide solution. The value so
obtained was 0.432 mol l^{-1}. An aliquot of 4.07 ml is equiva
to

$$\frac{0.432 \times 4.07}{0.3700} = 4.75 \text{ ml of HCl}$$

Since one molecule of n-pentyl fluoride removes one molecu
of sodium ethoxide, the *difference* between the two concentrations remains constant at (5.93 - 4.75) = 1.18 ml of 0.3700
HCl per 4.07 ml aliquot. The concentration of n-pentyl fluori
($b - x$) shown in *Table 13.4* is therefore obtained by subtract
1.18 from the corresponding sodium ethoxide concentration.

The differential rate law for unequal concentrations (equatio
2.17) is

$$\frac{dx}{dt} = k(a - x)(b - x)$$

The integrated form (equation 2.22) is

$$kt = \frac{1}{(a - b)} \ln\left[\frac{b(a - x)}{a(b - x)}\right]$$

The calculation of the rate constant from this equation is set out in *Table 13.4*. Since the logarithmic bracket $[b(a - x)/a(b - x)]$ contains only ratios of concentrations, the values of $(a - x)$ and $(b - x)$ can be expressed in ml of 0.3700 M HCl. However, the term $1/(a - b)$ has dimensions of $(concentration)^{-1}$, and so $(a - b)$ must be expressed in mol l^{-1} to give the second-order rate constant (k) with dimensions of $1\ mol^{-1}\ s^{-1}$. The rate constant is obtained by the multiplication of the logarithmic term by $2.303/(a - b)t$ (see M17). Thus the first value of the rate constant is

$$k = \frac{2.303 \times 0.00915}{(0.539 - 0.432) \times 1.59 \times 10^4} = 1.24 \times 10^{-5}\ 1\ mol^{-1}\ s^{-1}$$

The mean value of k (uncorrected for solvent expansion) is $1.26 \times 10^{-5}\ 1\ mol^{-1}\ s^{-1}$ at 120.0 °C. The solutions were, however, made up at 20 °C, and the 100 °C rise in temperature causes the ethanol to expand by 14%. The actual concentrations are therefore 14% lower than the nominal values. This does not affect the ratios in the logarithmic bracket, but the term $1/(a - b)$ increases by 14% (i.e. by a factor of 1.14). The corrected rate constant is therefore $1.26 \times 10^{-5} \times 1.14 = 1.44 \times 10^{-5}\ 1\ mol^{-1}$ s^{-1}. The calculation of the rate constant of this reaction by the method of least squares is shown in Section 14.2.

13.5 Calculation of Energy of Activation from Kinetic Measurements at Several Temperatures

The decomposition of benzenediazonium chloride in water, discussed in Section 13.1, has been carefully studied at several temperatures[*]. The experimental results shown in *Table 13.5* were obtained by measuring the change in pressure due to the evolution of nitrogen.

Table 13.5

Temperature / °C	k / s^{-1}	T/K	10^3 K/T	$-\log(k/s^{-1})$
15.1	9.30×10^{-6}	288.2	3.470	5.032
19.9	2.01×10^{-5}	293.0	3.413	4.697
24.7	4.35×10^{-5}	297.8	3.358	4.362
30.0	9.92×10^{-5}	303.1	3.300	4.003
35.0	2.07×10^{-4}	308.1	3.245	3.684
40.1	4.28×10^{-4}	313.2	3.193	3.369
44.9	8.18×10^{-4}	318.0	3.145	3.087

[*]Moelwyn-Hughes, E.A. and Johnson, P., *Trans. Faraday Soc.*, **36**, 954 (1940)

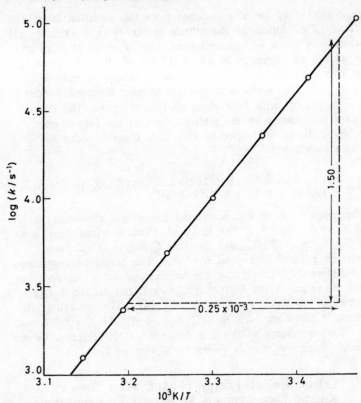

Figure 13.5 Energy of activation for decomposition of benzenediazonium chloride

The activation energy is calculated from these data using the Arrhenius equation 4.9

$$\ln k = \frac{-E^{\ddagger}}{RT} + \text{constant}$$

Using M3, a plot of $\ln k$ against $1/T$ will be a straight line of slope $-E/R$. From M17, if common logarithms are used the slope will be $-E/2.303R$.

The data in *Table 13.5* are plotted in *Figure 13.5*. From the graph

$$\text{slope} = \frac{1.50}{0.25 \times 10^{-3}} = 6.00 \times 10^3 \text{ K} = \frac{-E}{2.303R}$$

Taking $R = 8.314$ J K^{-1} mol^{-1}

$$E = 2.303 \times 8.314 \times 6.00 \times 10^3 = 114 \text{ kJ mol}^{-1}$$

13.6 Calculation of a Second-order Rate Constant using Potentiometric Data

In an aqueous solution containing a constant excess of bromide ions, the redox potential E of a smooth platinum electrode varies linearly with the logarithm of the concentration of bromine. For such an electrode, the Nernst equation is

$$E = E^{\ominus} + \frac{RT}{2F} \ln \frac{[Br_2]}{[Br^-]^2} = \left\{ E^{\ominus} - \frac{RT}{T} \ln [Br^-] \right\} + \frac{RT}{2F} \ln [Br_2]$$

Thus

$$E = E' + \frac{RT}{2F} \ln [Br_2] \tag{13.1}$$

where E' is a parameter whose value depends on the temperature and the concentration of bromide ions, both of which are kept constant in the kinetic experiment.

Suppose that one molecule of an aromatic substance ArH reacts with one molecule of bromine

$$ArH + Br_2 \rightarrow ArBr + HBr \tag{13.2}$$

Assuming that 13.2 is a second-order reaction

$$-\frac{d[Br_2]}{dt} = k[ArH][Br_2] \tag{13.3}$$

Rearranging 13.3 and using M21

$$-\frac{1}{[Br_2]} \frac{d[Br_2]}{dt} = -\frac{d \ln [Br_2]}{dt} = k[ArH] \tag{13.4}$$

By differentiating 13.1

$$\frac{dE}{dt} = \frac{RT}{2F} \frac{d \ln [Br_2]}{dt} \tag{13.5}$$

Putting 13.4 into 13.5

$$\frac{dE}{dt} = -k[ArH] \left(\frac{RT}{2F} \right) \tag{13.6}$$

At 25 °C, $2.303RT/F$ = 0.05915 V. Equation 13.6 then becomes at this temperature

$$k[\text{ArH}] = \frac{-2F}{RT}\frac{dE}{dt} = -77.8\frac{dE}{dt} \tag{13.7}$$

Equation 13.7 shows that providing [ArH] is kept constant, then dE/dt is constant, and so a straight line plot should be obtained for electrode potential versus time. From the slope of this line, which will be the same as the slope of the observed electromotive force E versus time, the rate constant can be calculated.

Typical results[*] obtained by this method are shown in *Figure 13.6*. The ordinate represents the e.m.f. of the cell obtained by combining the platinum redox electrode with a saturated calomel electrode (IUPAC potential of the SCE at 25 °C is +0.245 V). If the potential of the reference electrode and any junction potentials that may be present remain constant, there is no need to know their actual values. This is because the rate constant depends only on the slope of the graph of the e.m.f. of the cell against time.

In the example shown in *Figure 13.6* the slope is

$$\frac{dE}{dt} = -\frac{0.0711}{4} = -1.78 \times 10^{-2} \text{ V min}^{-1}$$

Hence from 13.7, at a constant concentration of ArH

$$k[\text{ArH}] = (-1.78 \times 10^{-2}) \times (-77.8) = 1.39 \text{ min}^{-1}$$

In principle, it is a simple matter to find the second-order rate constant by dividing by [ArH]. In the example quoted, however, ArH was NN-dimethylaniline and the solution was strongly acidic (a typical solution being 2 M in sulphuric acid). Under these circumstances most of the tertiary amine is present as the quaternary salt which is positively charged and does not react readily with bromine. The observed value of dE/dt therefore varies with the acid strength of the solution. For further details the reader should consult the work of Bell and Ramsden.

13.7 Calculation of the Volume of Activation

The hydrolysis of acetamide was investigated in a high-pressure reaction vessel under conditions of constant temperature and

[*]Bell, R.P. and Ramsden, E.N., *J. Chem. Soc.*, 162 (1958)

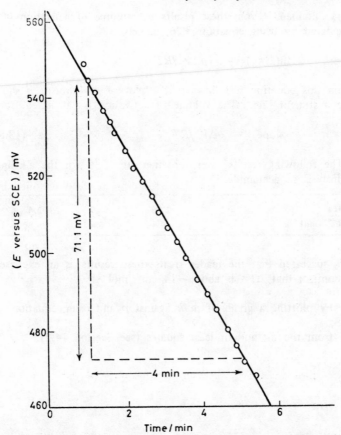

Figure 13.6 Potentiometric study of the bromination of NN-dimethyl-aniline

constant applied pressure. The progress of the reaction was followed by a sampling method using a colorimetric analytical procedure.

The second-order rate constant for the reaction

$$CH_3CONH_2 + OH^- \rightarrow CH_3CO_2^- + NH_3$$

was determined from the type of plot described in Section 13.3, i.e. a graph was plotted of $x/(a - x)$ against time. Several values of the rate constant at different constant applied pressures were obtained. From these results the volume of activation is calculated by using equation 5.26, namely

$$\ln(k/k_0) = -p\Delta V^{\ddagger}/RT$$

From this equation it follows that a plot of ln k versus p should give a straight line. The volume of activation is calculated from

$$\text{slope} = -\Delta V^{\ddagger}/RT \qquad (13.8)$$

The following results[*] were obtained at 25 °C for the alkaline hydrolysis of acetamide:

p/MPa	0.1	27.6	68.9	103.4
$k/10^{-5}$ l mol^{-1} s^{-1}	3.77	4.44	5.47	6.80

It is suggested that the reader treats these results as an exercise to confirm that $\Delta V^{\ddagger}$ is about -14 cm^3 mol^{-1}:

(i) by plotting a graph of ln k against p and using equation 13.8;
(ii) from the method of least squares (see Section 14.2).

[*]Laidler, K.J. and Chen, D., *Trans. Faraday Soc.*, 54, 1026 (1958). 1 MPa $\approx$ 10 atmospheres.

14

MATHEMATICS AND THERMODYNAMICS

14.1 Useful Mathematics

This section summarizes the mathematical results that have been assumed in derivations and calculations in the preceding chapters.

M1 If x is proportional to y ($x \propto y$), then

$$x = cy, \text{ where } c \text{ is a constant}$$

M2 The numerical value of the slope of a straight line graph of y against x is obtained by choosing two convenient points on the graph, and measuring the differences in values of y and x between the two points, i.e. Δy and Δx. These values must be expressed in the units in which the graph is plotted. The slope is then equal to $\Delta y / \Delta x$.

If the two points have co-ordinates (x_1, y_1) and (x_2, y_2), where x_2 is greater than x_1, then the slope equals $(y_2 - y_1)/(x_2 - x_1)$, which can be either positive or negative. This procedure is illustrated in *Figures 13.2, 13.3, 13.4* and *13.6*, in which time has been plotted along the x-axis.

M3 If $y = mx + c$ (where m and c are constants) a graph of y against x will be a straight line of slope m.

Straight line graphs can often be obtained from more complex expressions by correct choice of the variables to be plotted against one another. Thus if

$$F(x) = \text{a function of } x$$

and

$$G(y) = \text{a function of } y$$

then if

$$G(y) = mF(x) + c$$

a graph of $G(y)$ plotted against $F(x)$ will also be a straight line of slope m. This is illustrated in *Figure 13.5* where $G(y)$ is $-\log k$ and $F(x)$ is $10^3/T$.

M4 $$X^a \times X^b = X^{a+b}$$

M5 $$X^a \div X^b = X^{a-b}$$

M6 From M5, $X^0 = X^{a-a} = X^a \div X^a = 1$

M7 Factorial X is written as $X!$ As an example

$$8! = 1 \times 2 \times 3 \times 4 \times 5 \times 6 \times 7 \times 8$$
$$n! = 1 \times 2 \times 3 \times 4 \times \ldots \times n$$

It can be shown that factorials correspond to the mathematical 'gamma function' of whole numbers, and hence that $0! = 1$.

M8 The exponential e (the base of natural logarithms) is a number equal to 2.7183 (to four places of decimals). e raised to the power of x (e^x) is denoted by exp x. The exponential function may be regarded as the inverse of the logarithmic function.

M9 Analogous to M4 and M5 are the equations

$$\exp(a) \times \exp(b) = \exp(a + b)$$
$$\exp(a) \div \exp(b) = \exp(a - b)$$

M10 Natural logarithms are closely related to the exponential function. The natural logarithm of x is written ln x. If

$$x = \exp(y)$$

then

$$\ln x = y$$

M11 The mutual relationship of the logarithmic and exponential function is shown by the equation

$$\exp(\ln x) = x = \ln(\exp x)$$

M12   In common logarithms the exponential number e is replaced by the number ten. Thus if

$$y = \log x, \text{ then } x = 10^y$$

M13 $$\ln x + \ln y = \ln(xy)$$

and

$$\log x + \log y = \log(xy)$$

M14 $$\ln x - \ln y = \ln(x/y)$$

and

$$\log x - \log y = \log(x/y)$$

M15   If $\ln x = y + c$, where c is a constant then from M10, $x = \exp(y + c)$;

Using M9, $x = \exp y \times \exp c$
Since c is a constant, $\exp c$ is also a constant, equal to C, say.
Therefore, $x = C \exp y$.

M16   If y is proportional to x, and t is a third variable, then the slope of the graph of $\ln y$ against t is equal to that of $\ln x$ against t.

For, by M1, $y = cx$

Using M13, $\ln y = \ln c + \ln x$
Since c is a constant, $\ln c$ is a constant.
Therefore $\ln y$ and $\ln x$ differ only by a constant which does not affect the slope of the graph. This means that absolute values of y are not required to find the slope of a graph of $\ln y$. All that is needed is a quantity proportional to y. This result has been used in Section 13.2.

M17   Natural and common logarithms are related by the equation

$$\ln x = \ln 10 \times \log x$$

The numerical value of $\ln 10$ is 2.303
Hence $\ln x = 2.303 \log x$

M18 The result of differentiating y with respect to x is written as dy/dx. The physical significance of dy/dx is that it represents the slope of the tangent to the curve obtained by plotting a graph of y against x. If a small increase in x produces an increase in y, the curve will have a positive slope and dy/dx will be positive. On the other hand, if a small increase in x causes a decrease in y, dy/dx will be negative.

M19 Partial differentiation refers to a system of more than two variables. The symbol $\partial V/\partial T$ has the same basic meaning as dV/dT but contains the message that there are other variables apart from V and T that affect the system. The other variables may be specified explicitly by placing them as a subscript to the differential expression. For example

$$\left(\frac{\partial V}{\partial T}\right)_p$$

refers to the slope of the curve of V against T with the restriction that p is constant.

Since experimental conditions such as constant temperature, pressure, volume, concentration, etc. are often imposed on experiments, partial differentiation is frequently encountered in scientific work, particularly in thermodynamics.

M20 If

$$y = x^n$$

$$dy/dx = nx^{n-1}$$

M21 Integration is the reverse of differentiation. The two integrals used in this book are:

$$\int x^n \, dx = \frac{x^{n+1}}{n+1} + I \text{ (when } n \text{ is not equal to } -1)$$

where I is a constant called the constant of integration. Thus

$$\int \frac{dx}{x^2} = -\frac{1}{x} + I \quad \text{and} \quad \int \frac{dx}{(a-x)^2} = \frac{1}{a-x} + I$$

M22

$$\int dx/x = \ln x + I$$

It is this integral that gives rise to the frequent occurrence of natural logarithms in physical chemistry.

M23 The area under the curve of y plotted against x, between the limits of x_1 and x_2, is

$$\int_{x_1}^{x_2} y \, dx$$

14.2 Linear Least-squares Method

In interpreting experimental results it is often necessary to find the best straight line through a set of observed points. This can be done by eye after plotting the points on a graph, but more reliable and reproducible estimates can be obtained by using a statistical technique, such as the *method of least squares.* With the increasing availability and range of facilities of electronic calculators, the least-squares estimate of a slope should become a commonplace technique.

The criterion defining the best line is that the sum of the squares of the deviations of the observed points from the line shall be a minimum. This, however, is not a unique definition as deviations could be measured parallel to the y-axis, parallel to the x-axis or as perpendicular distances from the line.

If there is reason to believe that the errors in the observed value are associated with a dependent variable (plotted along the y-axis) and that the values of the independent variable (plotted along the x-axis) are error free, then the least-squares line through the points is

$$y = a + bx \tag{14.1}$$

where the intercept (a) and the slope (b) are defined by

$$\Sigma y = na + b\Sigma x \tag{14.2}$$

$$\Sigma xy = a\Sigma x + b\Sigma x^2 \tag{14.3}$$

n being the number of points through which the line is being fitted. The operator Σ denotes 'the sum of', e.g. Σx is the sum of the n values of x.

Many electronic calculators have the facility for calculating Σx and Σx^2 at the same time, which greatly speeds the calculation. Similarly, if it is possible to calculate Σx at the same time as Σxy a useful calculating check can be introduced.

It should be noted from equation 14.3 that if $\Sigma x = 0$ or if $a = 0$, the slope of the least-squares lines is

$$b = \frac{\Sigma xy}{\Sigma x^2} \tag{14.4}$$

Thus equation 14.4 gives the slope of the best line through the origin ($a = 0$).

In using these formulae, it is essential to choose as the independent variable (x) the basic property which is known with some accuracy. In chemical kinetics this is often time or reciprocal temperature.

To apply equations 14.2 and 14.3, values of Σx, Σy, Σxy and Σx^2 are calculated (and checked) for the n points through which the line is to be fitted. The calculated values are substituted in equations 14.2 and 14.3 which will result in two simultaneous equations in a and b, which are solved in the usual way. The method of least squares is applied below to the reaction of n-pentyl fluoride with sodium ethoxide, using data taken from *Table 13.4*, p. 158.

The independent variable (x) is time, values of which are shown in the first column of the table. The dependent variable (y) is the log term, shown in the fifth column of the table. The unit of time in the table corresponds to 10^4 s. Using a calculator:

$$
\begin{aligned}
\Sigma x &= 58.81 \\
\Sigma y &= 0.339\,99 \\
\Sigma x^2 &= 681.8563 \\
\Sigma xy &= 3.930\,074\,7
\end{aligned}
$$

The value of n, the number of points, is 8, because the origin counts as a point. This is because at zero time there should have been no reaction. From equations 14.2 and 14.3

$$
\begin{aligned}
0.339\,99 &= 8a + 58.81b \\
3.930\,074\,7 &= 58.81a + 681.8563b
\end{aligned}
$$

Solving for b by multiplying the first equation by 58.81/8 and subtracting*

$$b = 0.005\,734 \quad \text{(in units of } 10^4 \text{ s); } a = 0.000\,345\,2$$

*a and b above should not be confused with concentrations.

From equation 2.22 and M17, the value of the least-squares slope is

$$b = (0.539 - 0.432)k/2.303$$

Hence

$$k = \frac{2.303 \times 0.005\ 734 \times 10^{-4}}{(0.539 - 0.432)} = 1.234 \times 10^{-5} \text{ l mol}^{-1} \text{ s}^{-1}$$

in agreement with the values in the sixth column of *Table 13.4.*

14.3 Outline of Thermodynamic Equilibria

The idea of a chemical equilibrium, which is fundamental to the theory of reaction rates, can be approached from two completely different viewpoints.

Firstly, there is the *dynamic* or *kinetic* approach as used in the early kinetic investigations on the hydrolysis of ethyl acetate (p. 2). Reactions are regarded as being chemically reversible, at least in principle, so that the equilibrium state can be defined as the one in which the (finite)* rates of the forward and reverse reactions are equal.

The rates of chemical reactions are themselves governed by rate laws depending on the concentrations of reactants, and so this kinetic view leads to the idea that the position of equilibrium should be definable in terms of the *concentrations* of reactants and products. Equilibrium constants calculated in this way are referred to as *classical* or *formal equilibrium constants.*

Secondly, there is the *thermodynamic* approach, developed from the work of J. Willard Gibbs, in which a system is said to be in an equilibrium state if any small displacement from that state leads to an increase in energy or a decrease in entropy. From this viewpoint, kinetic features such as the order of the reaction are completely irrelevant to the formulation of the conditions of equilibrium, so the concentrations of reactants and products are not directly involved.

One of the great concepts that arose from Gibbs' thermodynamic investigations was that a *potential* can be associated with *chemical*

---

*It is necessary to specify 'finite' here as a mixture of hydrogen and oxygen at STP in the absence of catalyst or spark produces equal rates of the forward and reverse reactions (namely a zero rate). This is not a thermodynamic equilibrium (see Section 11.9).

energy just as a potential can be associated with gravitational, electrical or magnetic energy. One common feature of any potential is that it cannot be measured absolutely. All that can be done is to set up some standard state and measure *potential differences* (e.g. with gravitational energy, sea-level is arbitrarily defined as having zero potential).

Chemical potential (μ) can be measured as the quantity of Gibbs' free energy associated with one mole of substance under the experimental conditions. Chemical potential varies with temperature, pressure and concentration in accordance with the usual thermodynamic relationships. The value of μ (and that of any property derived from it) also depends on the value chosen for the standard chemical potential (μ°). The significance of this simple but far-reaching conclusion is sometimes overlooked.

Suppose that the chemical system described by

$$aA + bB \rightleftharpoons pP + qQ \tag{14.5}$$

has reached a state of thermodynamic equilibrium at constant temperature and pressure. Then the Gibbs free energy change (ΔG) corresponding to transforming a moles of A, etc. into products is zero (a much larger reaction mass is assumed so that concentrations are not significantly affected). The reaction has no additional driving force and it is evident that the reactance and product are balanced by the condition

$$(a\mu_A + b\mu_B) = (p\mu_P + q\mu_Q)$$

Thus

$$\Delta G = (p\mu_P + q\mu_Q +) - (a\mu_A + b\mu_B +) = 0 \tag{14.6}$$

All that is needed to obtain the expression for the thermodynamic equilibrium constant from equation 14.6 is to relate the chemical potential to the *activity* (a) of each substance involved in the equilibrium 14.5. This is done by means of the *definition*

$$\mu - \mu^{\circ} = RT \ln a \tag{14.7}$$

The reason for defining activity in terms of chemical potential differences in this way is that it can easily be shown that the relation of chemical potential to pressure for a single ideal gas is

$$\mu - \mu^{\ominus} = RT \ln(p/p^{\ominus})$$ (14.8)

Hence for an ideal gas the activity is a ratio of the pressure (p) to some standard pressure $(p^{\ominus})$.

If the definition of activity in 14.7 is substituted into the condition for thermodynamic equilibrium 14.6 then

$$p(\mu_P^{\ominus} + RT \ln a_P) + q(\mu_Q^{\ominus} + RT \ln a_Q)$$
$$-a(\mu_A^{\ominus} + RT \ln a_A) - b(\mu_B^{\ominus} + RT \ln a_B) = 0$$ (14.9)

Collecting terms, using M13 and M14

$$p\mu_P^{\ominus} + q\mu_Q^{\ominus} - a\mu_A^{\ominus} - b\mu_B^{\ominus} = \Delta G^{\ominus} = -RT \ln \frac{a_P^p \, a_Q^q}{a_A^a \, a_B^b}$$ (14.10)

The right-hand side of 14.10 can be written in the form sometimes known as the *van't Hoff isotherm* by making the substitution

$$K = \frac{a_P^p \, a_Q^q}{a_A^a \, a_B^b}$$ (14.11)

so that

$$\Delta G^{\ominus} = -RT \ln K$$ (14.12)

It will be seen that the thermodynamic equilibrium constant K (defined by equation 14.11) is the same type of expression as the formal equilibrium constant

$$K_c = \frac{[P]^p [Q]^q}{[A]^a [B]^b}$$ (14.13)

It should be noted that whereas the formal equilibrium constant has dimensions of (concentration)$^{p+q-a-b}$, the thermodynamic equilibrium constant is always dimensionless as it is defined in terms of activities which are themselves dimensionless. However, this difference in dimensions between the thermodynamic and formal equilibrium constants can be eliminated if each concentration term is made a ratio of a chosen standard state.

For solutions, a convenient standard state is an idealized hypothetical solution of unit concentration, because this definition leads to the result that in dilute solution, in which the solute behaves ideally, the activity of the solute is proportional to its concentration.

In general, the ratio of activity to concentration is called the *activity coefficient* (*y*). Hence by combining 14.11 and 14.13

$$K = K_c \frac{y_P^p \, y_Q^q}{y_A^a \, y_B^b} \tag{14.14}$$

where the *y*'s are the appropriate activity coefficients. Equation 14.14 is the basis of the analysis of kinetic salt effects.

SUGGESTIONS FOR ADDITIONAL READING

1. ASHMORE, P.G., *Principles of Reaction Kinetics*, Chemical Society Monographs for Teachers, Number 9, Second Edition, The Chemical Society, London (1973)
2. AVERY, H.E., *Basic Reaction Kinetics and Mechanisms*, Macmillan, London (1974)
3. BELL, R.P., *The Proton in Chemistry*, Chapman and Hall, London, Second Edition (1973)
4. BERNHARD, S. *The Structure and Function of Enzymes*, Benjamin, Menlo Park, Calif. and London (1968)
5. BOND, G.C., *Principles of Catalysis*, Chemical Society Monographs for Teachers, Number 7, Second Edition, The Chemical Society, London (1972)
6. BRADLEY, J.N., *Fast Reactions,* Clarendon, Oxford (1975)
7. BUNKER, D.L., *Theory of Elementary Gas Reactions*, Pergamon, Oxford (1966)
8. CALDIN, E.F., *Fast Reactions in Solution*, Blackwell, Oxford (1964)
9. GARDINER, W.C., *Rates and Mechanisms of Chemical Reactions*, Benjamin, Menlo Park, Calif. and London (1969)
10. HAMMES, G.G., *Principles of Chemical Kinetics,* Academic, London and New York (1978)
11. LAIDLER, K.J. and BUNTING, P.S. *The Chemical Kinetics of Enzyme Action*, Clarendon, Oxford, Second Edition (1973)
12. MARK, H.B. and RECHNITZ, G.A., *Kinetics in Analytical Chemistry*, Interscience, New York and Chichester (1972)
13. MOELWYN-HUGHES, E.A., *Chemical Statics and Kinetics of Solution*, Academic, London and New York (1971)
14. NICHOLAS, J., *Chemical Kinetics: a Modern Survey of Gas Reactions*, Harper and Row, London and New York (1976)
15. PILLING, M.J., *Reaction Kinetics,* Clarendon, Oxford (1975)
16. ROBINSON, P.J. and HOLBROOK, K.A., *Unimolecular Reactions,* Wiley, New York and Chichester (1972)
17. SCRIMGEOUR, K.G., *Chemistry and Control of Enzyme Reactions,* Academic, London and New York (1977)
18. WESTON, R.E. and SCHWARTZ, H.A., *Chemical Kinetics*, Prentice-Hall, Englewood Cliffs, N.J. and Hemel Hempstead (1972)

INDEX